LAMINAR MOTION OF MULTIPHASE MEDIA IN CONDUITS

LAMINARNOE DVIZHENIE MNOGOFAZNYKH SRED V TRUBOPROVODAKH

ЛАМИНАРНОЕ ДВИЖЕНИЕ МНОГОФАЗНЫХ СРЕД В ТРУБОПРОВОДАХ

LAMINAR MOTION OF MULTIPHASE MEDIA IN CONDUITS

Dzharulla F. Faizullaev

Scientific Director
Institute of Mechanics and Earthquakeproof Construction
Academy of Sciences of the Uzbek SSR
Tashkent, Uzbek SSR

Translated from Russian

Springer Science+Business Media, LLC 1969

Dzharulla Faizullaevich Faizullaev was born in Tashkent in 1924. A graduate of Central Asia (now Tashkent) State University, he was, until 1960, an instructor in the higher mathematics department of the Tashkent Institute of Irrigation and Agricultural Mechanization Engineers. Since 1960 he has been associated with the Institute of Mechanics and Earthquakeproof Construction (formerly the Mechanics Institute) of the Academy of Sciences of the Uzbek SSR, where he organized and presently heads the hydrodynamics laboratory. He was director of the Mechanics Institute in 1962-1963 and has been scientific director of the Institute of Mechanics and Earthquakeproof Construction since April 1966.

Library of Congress Catalog Card Number 69-12511

The Russian text, published for the Institute of Mechanics of the Uzbek SSR by FAN Press in Tashkent in 1966, has been corrected by the author for the English edition.

Джарулла Файзуллаевич Файзуллаев

ЛАМИНАРНОЕ ДВИЖЕНИЕ МНОГОФАЗНЫХ СРЕД В ТРУБОПРОВОДАХ

ISBN 978-1-4899-4832-8   ISBN 978-1-4899-4830-4 (eBook)
DOI 10.1007/978-1-4899-4830-4

© 1969 Springer Science+Business Media New York

Originally published by Consultants Bureau in 1969.

Softcover reprint of the hardcover 1st edition 1969

All rights reserved

No part of this publication may be reproduced in any form without written permission from the publisher

CONTENTS

INTRODUCTION

A study of the phenomena occurring in the motion of multiphase media is essential for the solution of many theoretical and engineering problems: the theory of wave propagation in water-saturated ground and dust-saturated air, the theory of fine grinding, the theory of mass entrainment from the surface of a body by a high-parameter flow, the theory of the flow around porous bodies, the theory of the motion of powder gases containing unburned particles, the perfection of water and pneumatic transport, petroleum output and petroleum refining when there is foreign matter in the petroleum, etc. In all these cases the moving material is a mixture of two or more media, i.e., phases.

A great deal of experimental material on multiphase, particularly two-phase, media has recently been collected, and a number of theories have been constructed on the basis of these data. The theory of the motion of two-phase media has developed in three principal directions.

1. The mixture is considered a single-phase fluid with averaged motion parameters (stream velocity, coefficient of viscosity of the mixture, etc.) The mixture is considered a Newtonian fluid, i.e., one in which the stress is proportional to the shear strain rate:

$$\tau_{mi} = \mu_{mi}\, \frac{\partial u_{mi}}{\partial n},$$

where μ_{mi} is the coefficient of viscosity of the mixture. In this case the influence of the interaction of the media on the motion is ignored.

2. In the second approach, which was developed after the first, the mixture is not a Newtonian fluid. It obeys Bingham's law

$$\tau = \mu \frac{\partial u}{\partial n} + \tau_0,$$

where τ_0 is the initial tangential stress. Such mixtures are called viscoplastic fluids and theories for them have been developed extensively. However, in some cases viscoplastic flow in conduits bounds a two-phase flow of Newtonian fluids. The connection between the two-phase flow parameters in such a motion and the viscoplastic flow parameters has not been examined up to now. A joint investigation of these two flows can be conducted on the basis of the theory advanced by Rakhmatulin [1].

3. The third approach, which is fairly new, differs from the others in that the relative phase velocities are taken into account. This as well as the first approach has been developed experimentally and theoretically for two modes of motion: laminar and turbulent. A general formulation of the theory of laminar interpenetrating motion (in the presence of a relative velocity) is given in [1]. Each medium is considered continuous, and its motion is examined as motion in the moving and changing porous medium formed by the other phases (media). The

interaction of the media due to changes in the stream tube cross section, as well as to friction between the media, which is proportional to the relative velocity, is taken into account.

The system of equations describing the laminar motion of a mixture, without using any empirical relationships, is a closed system of n equations in n unknowns. This permits theoretical study of many phenomena occurring in a mixture in motion. Since to date there exist no measuring instruments or apparatus which could be used to study all mixture motions experimentally, theoretical results take on particular importance.

Rakhmatulin's paper [1] allows us to study diverse motions of mixtures consisting of any number of both incompressible and compressible* and ideal as well as viscous media. By analogous reasoning, equations of motion may be formulated for mixtures in which one of the media is viscoplastic while the others are Newtonian fluids.

It follows that by starting from the theory of interpenetrating motions of multiphase media, diverse problems can be solved and many already have been solved. To some degree the present work generalizes almost all the work in which the interpenetrating motion of mixtures in conduits has been considered. The differential equations of motion and the continuity equations are given in Chapter I, and solutions of the problems of the motion of ideal two- and three-phase media are presented in Chapter II (all the media taking part in the motion are here considered ideal).

Steady-state motions of viscous and viscous–ideal multiphase media in finite and infinite conduits are explored in Chapter III for constant media porosity. The motion of a multiphase mixture will be steady if the motion of each phase taking part in the motion of the mixture is steady. Note that because of the interaction between phases it is not possible for the motion of one of the phases to be steady while that of the others is unsteady.

Two cases of mixture motion are considered in this chapter. First, each phase of the mixture is considered viscous; this case is also applicable to computations of the motion of a mixture consisting of a viscous fluid and very fine suspended solid particles. It is known that the coefficient of viscosity of a mixture of viscous fluids differs from the coefficients of viscosity of the separate phases. However, the coefficient of viscosity of the mixture may be expressed in terms of the coefficients of viscosity of the individual phases.

If very fine solid particles are contained in a viscous fluid, then just as before, the coefficient of viscosity of the mixture will differ from the coefficient of viscosity of the liquid phase. Thanks to the introduction of the concept of reduced density in [1], each phase can be considered a continuum. Therefore, by considering the solid phase as a continuum, and treating it provisionally as a Newtonian fluid, we introduce the coefficient of viscosity of the solid phase, which we call the "provisional coefficient of viscosity." Note that this coefficient is valid only when the solid particles are contained in a fluid.

The second case of Chapter III deals with the motion of a viscous–ideal mixture. Here one phase of the two-phase mixture is viscous and the other is ideal. This case is also extended to the motion of gas–liquid mixtures. Let us note that in all the cases listed the media possess the effect of viscosity (even if they are considered ideal), which is taken into account by the interaction coefficient K.

In Chapter IV attention is paid to the unsteady motions of viscous two-phase media in conduits; particular problems are solved here, since the porosities are considered constant and independent of the time or the coordinates.

---

* Only barotropic fluids are considered at present.

The motions of viscous two-phase media when the porosities are variable are examined in Chapter V. The problem is to find the values of the motion parameters in any cross section when their values are given in some other cross section (considered to be the initial section) of a conduit extending to infinity in both directions. We may not take arbitrary values of the parameters in the initial section, but they must satisfy the generalized Poiseuille equations for multiphase media.

A computation of the drag coefficient of a conduit for the motion of two-phase media is given in Chapter VI.

Various methods of determining the interaction coefficients and the provisional coefficient of viscosity are elucidated in Chapter VII, the results of experimental investigations laying the foundation for the existence of laminar flow of two-phase media (water – rosin) are presented, and the values of the above-mentioned coefficients are found.

Throughout, whenever the motion of a mixture is examined, it is assumed that each phase (liquid or solid) is uniformly distributed over the cross section of the conduit.

We hope that the present monograph will contribute, to some degree, to the solution of specific engineering problems.

Critical comments, for which we express our gratitude in advance, should be sent to the Mechanics Institute of the Academy of Sciences of the UzbekSSR.

The author expresses his heartfelt gratitude to Academicians K. A. Rakhmatulin, M. T. Urazbaev, and V. K. Kabulov, and also to D. F. Shul'gin and P. A. Tsitovich for valuable comments.

CHAPTER I

DIFFERENTIAL EQUATIONS OF THE MOTION
OF MUTUALLY PENETRATING MULTIPHASE MEDIA

§1. Theorem on the Equality of the Volume and Surface

Porosities in a Homogeneously Porous Medium

Let a unit volume contain a pore volume of f_ω ; we shall call this the volume porosity.

Let f_s denote the area of the pore per unit surface of a given porous medium. We call this number the surface porosity. Let us prove the following theorem: The volume and surface porosities in a homogeneously porous medium are equal, i. e., $f_\omega = f_s$.

Let some fluid with the true density ρ_i move in a porous medium. The reduced density of this fluid ρ_c, obtained for a uniform distribution of its mass over the whole volume, will be, in the absence of a porous medium;

$$\rho_c = f_\omega \, \rho_i. \tag{1.1}$$

Motion in a porous medium with velocity V and density ρ_i is identical to the motion in a free medium with velocity V and density ρ_c. The discharge through unit surface and the true density equals $\rho_i V_n f_s$, and through the reduced density $\rho_c V_n$. Equating these expressions, we obtain

$$\rho_c = f_s \, \rho_i.$$

Substituting this expression into (1.1), we have

$$f_\omega = f_c \, .$$

§2. Application of the Theorem

The theorem we have just proved is quite important for taking account of the volume and surface forces in forming the ref equations of motion [1]. For example, if a pressure p acts on the area $d\sigma$, the force per fraction of the n-th medium in the multiphase mixture will be

$$p f_n \, d\sigma = \frac{\rho_{nc}}{\rho_{ni}} p d\sigma,$$

where f_n is the porosity for the n-th phase of the mixture.

Since we have for the n media on the basis of the porosity definition

$$f_1 + f_2 + \cdots + f_n = 1,$$

4

then, by virtue of (1.1), we obtain a relationship needed subsequently :

$$\frac{\rho_1}{\rho_{1i}} + \frac{\rho_2}{\rho_{2i}} + \cdots + \frac{\rho_n}{\rho_{ni}} = 1 \tag{2.1}$$

where ρ_n and ρ_{ni} are the reduced and true densities of the n-th medium.

§ 3. Lomonosov's Law (Conservation of Matter) in Euler Form (Continuity Equation)

If the mixture consists of n media, then as is seen from the above, besides the n true densities there are still n reduced densities at each point of space, and the corresponding medium is continuous relative to each of them. Let us note that there is as much velocity at each point of the moving mixture as there is medium at the point. In that case Lomonosov's law (conservation of matter) is written in Euler form [1]

$$\frac{\partial \rho_n}{\partial t} + \operatorname{div}\left(\rho_n \, \vec{V} \right) = 0 \tag{3.1}$$

for the reduced densities ρ_n. Equation (3.1) becomes

in Cartesian coordinates

$$\frac{\partial \rho_n}{\partial t} + \frac{\partial (\rho_n u_n)}{\partial x} + \frac{\partial (\rho_n v_n)}{\partial y} + \frac{\partial (\rho_n w_n)}{\partial z} = 0, \tag{3.2}$$

in a cylindrical system

$$\frac{\partial \rho_n}{\partial t} + \frac{\partial (\rho_n v_{nr})}{\partial r} + \frac{1}{r} \frac{\partial (\rho_n v_{n\varphi})}{\partial \varphi} + \frac{\partial (\rho_n v_{nz})}{\partial z} + \frac{\rho_n v_{nr}}{r} = 0, \tag{3.3}$$

in a spherical system

$$\frac{\partial \rho_n}{\partial t} + \frac{\partial \rho_n v_{nr}}{\partial r} + \frac{1}{r} \frac{\partial (\rho_n v_{n\varphi})}{\partial \varphi} + \frac{1}{r \sin \varphi} \frac{\partial (\rho_n v_{n\psi})}{\partial \psi} + \frac{2\rho_n v_{nr}}{r} + \frac{\rho_n v_{n\varphi}\cot \varphi}{r} = 0,$$

$$n = 1 \ldots N, \tag{3.4}$$

where N is the number of phases in these formulas.

§ 4. Derivation of the Equations of Motion in a Cartesian Three-Dimensional Coordinate System

If an element of volume with sides dx, dy, dz is isolated from a multiphase mixture, then the sum of surface forces per n-th phase acting on dydz, say, will equal

$$\frac{\partial \left(\overline{p}_{nx} \, dydz \right)}{\partial x} dx,$$

where $\overline{p}_{nx}$ is a vector equal in absolute value to the stress vector on an area perpendicular to 0x axis, and directed oppositely. Besides this force there are interaction forces. The force acting on the n-th phase of the remaining phases consists of two components: (a) the forces $\overline{P}'_n$ dxdydz due to variability of the stream tube section (diffusor effect); the projections of this force on the coordinate axes are written as [1]

$$p dy \, dz \frac{\partial f_n}{\partial x} dx, \quad p dx \, dz \frac{\partial f_n}{\partial y} dy, \quad p dx \, dy \frac{\partial f_n}{\partial z} dz,$$

where

$$f_n = \frac{\rho_n}{\rho_{nl}}$$

is the porosity and p is the pressure, and (b) the forces due to the relative velocity of media motion P_n'' dxdydz, where

$$P_n' = \sum_{j=1}^{N} K_{jn}(\overline{V}_j - \overline{V}_n),$$

and $\overline{V}_n \overline{V}_j$ are the velocity vectors of the n-th and j-th phases, respectively; K_{jn} is the inter-action coefficient of the n-th and j-th phases. In this case the equation of motion for the n-th medium is represented as

$$\rho_n \frac{d\overline{v}_n}{dt} = \frac{\partial \overline{p}_{nx}}{\partial x} + \frac{\partial \overline{p}_{ny}}{\partial y} + \frac{\partial \overline{p}_{nz}}{\partial z} + \overline{P}_n' + \overline{P}_n'' + \overline{F}_n \rho_n, \qquad (4.1)$$

where the forces $\overline{P}_n'$, $\overline{P}_n''$ are referred to the volume element of the whole mixture; $\overline{F}_n$ is the mass force with components X_n, Y_n, Z_n acting on unit mass of the n-th phase.

When the equation of motion (4.1) is projected onto the coordinate axes, we obtain

$$\left. \begin{aligned}
\rho_n \frac{du_n}{\partial t} &= \frac{\partial p_{nxx}}{\partial x} + \frac{\partial p_{nyx}}{\partial y} + \frac{\partial p_{nzx}}{\partial z} + p\frac{\partial f_n}{\partial x} + \sum_{j=1}^{N} K_{jn}(u_j - u_n) + X_n \rho_n \\
\rho_n \frac{dv_n}{\partial t} &= \frac{\partial p_{nxy}}{\partial x} + \frac{\partial p_{nyy}}{\partial y} + \frac{\partial p_{nzy}}{\partial z} + p\frac{\partial f_n}{\partial y} + \sum_{j=1}^{N} K_{jn}(v_j - v_n) + Y_n \rho_n \\
\rho_n \frac{dw_n}{dt} &= \frac{\partial p_{nxz}}{\partial x} + \frac{\partial p_{nyz}}{\partial y} + \frac{\partial p_{nzz}}{\partial z} + p\frac{\partial f_n}{\partial z} + \sum_{j=1}^{N} K_{jn}(w_j - w_n) + Z_n \rho_n
\end{aligned} \right\} \qquad (4.2)$$

where u_n, v_n, w_n are the components of the velocity $\overline{V}_n$.

Let us assume that the generalized Newtonian hypothesis establishing a differential con-nection between the stress components and the velocities of the fluid particles is valid for each medium. However, the porosity must be taken into account in expressing such a connection. We then have

$$\left. \begin{aligned}
p_{nxx} &= f_n \left(-p - \frac{2}{3}\mu_n \operatorname{div}\overline{V}_n + 2\mu_n \frac{\partial u_n}{\partial x} \right) \\
p_{nyy} &= f_n \left(-p - \frac{2}{3}\mu_n \operatorname{div}\overline{V}_n + 2\mu_n \frac{\partial v_n}{\partial y} \right) \\
p_{nzz} &= f_n \left(-p - \frac{2}{3}\mu_n \operatorname{div}\overline{V}_n + 2\mu_n \frac{\partial w_n}{\partial z} \right) \\
p_{nxy} &= p_{nyx} = f_n \mu_n \left(\frac{\partial u_n}{\partial y} + \frac{\partial v_n}{\partial x} \right) \\
p_{nxz} &= p_{nzx} = f_n \mu_n \left(\frac{\partial u_n}{\partial z} + \frac{\partial w_n}{\partial x} \right) \\
p_{nyz} &= p_{nzy} = f_n \mu_n \left(\frac{\partial v_n}{\partial z} + \frac{\partial w_n}{\partial y} \right)
\end{aligned} \right\} \qquad (4.3)$$

where μ_n is the coefficient of viscosity of the n-th medium.

Substituting (4.3) into (4.2), we will have the equation of motion for the n-th phase taking part in the mixture motion, in three-dimensional Cartesian coordinates:

$$\rho_n \frac{du_n}{dt} = -f_n \frac{\partial p}{\partial x} + \frac{\partial}{\partial x} f_n \mu_n \left(2\frac{\partial u_n}{\partial x} - \frac{2}{3} \operatorname{div} \overline{V}_n\right) + \frac{\partial}{\partial y} f_n \mu_n \left(\frac{\partial v_n}{\partial x} + \frac{\partial u_n}{\partial y}\right) +$$

$$+ \frac{\partial}{\partial z} f_n \mu_n \left(\frac{\partial w_n}{\partial x} + \frac{\partial u_n}{\partial z}\right) + \sum_{j=1}^{N} K_{jn}(u_j - u_n) + X_n \rho_n$$

$$\rho_n \frac{dv_n}{dt} = -f_n \frac{\partial p}{\partial y} + \frac{\partial}{\partial y} f_n \mu_n \left(2\frac{\partial v_n}{\partial y} - \frac{2}{3} \operatorname{div} \overline{V}_n\right) + \frac{\partial}{\partial z} f_n \mu_n \left(\frac{\partial w_n}{\partial y} + \frac{\partial v_n}{\partial z}\right) +$$

$$+ \frac{\partial}{\partial x} f_n \mu_n \left(\frac{\partial u_n}{\partial y} + \frac{\partial v_n}{\partial x}\right) + \sum_{j=1}^{N} K_{jn}(v_j - v_n) + Y_n \rho_n$$

$$\rho_n \frac{dw_n}{dt} = -f_n \frac{\partial p}{\partial z} + \frac{\partial}{\partial z} f_n \mu_n \left(2\frac{\partial w_n}{\partial z} - \frac{2}{3} \operatorname{div} \overline{V}_n\right) + \frac{\partial}{\partial x} f_n \mu_n \left(\frac{\partial u_n}{\partial z} + \frac{\partial w_n}{\partial x}\right) +$$

$$+ \frac{\partial}{\partial y} f_n \mu_n \left(\frac{\partial v_n}{\partial z} + \frac{\partial w_n}{\partial y}\right) + \sum_{j=1}^{N} K_{jn}(w_j - w_n) + Z_n \rho_n$$

$$(n = 1 \ldots N)$$

$$(4.4)$$

Therefore, the system of equations (4.4), (3.2), (2.1) characterizes the motion of each phase of the mixture. In the case of constant true densities, this system will be closed, i.e., the number of equations will equal the number of unknowns. Let us note that the reduced densities in all the above equations are variable in the general case [1], even if the media of a two-phase flow are incompressible. Starting from this assumption, it is easy to note that although for an incompressible homogeneous fluid

$$\operatorname{div} \overline{V} = 0,$$

for an incompressible two-phase medium

$$\operatorname{div} \overline{V}_1 \neq 0, \quad \operatorname{div} \overline{V}_2 \neq 0.$$

These inequalities vanish if the reduced densities are constant.

§ 5. Equations of Motion for Viscous Three-Phase Mixtures

For viscous three-phase mixtures the system of equations (4.4) becomes

$$\rho_1 \frac{du_1}{dt} = -f_1 \frac{\partial p}{\partial x} + \frac{\partial}{\partial x} f_1 \mu_1 \left(2\frac{\partial u_1}{\partial x} - \frac{2}{3} \operatorname{div} \overline{V}_1\right) + \frac{\partial}{\partial y} f_1 \mu_1 \left(\frac{\partial v_1}{\partial x} + \frac{\partial u_1}{\partial y}\right) +$$

$$+ \frac{\partial}{\partial z} f_1 \mu_1 \left(\frac{\partial w_1}{\partial x} + \frac{\partial u_1}{\partial z}\right) + K_{12}(u_2 - u_1) + K_{13}(u_3 - u_1) + X_1 \rho_1$$

$$\rho_1 \frac{dv_1}{dt} = -f_1 \frac{\partial p}{\partial y} + \frac{\partial}{\partial y} f_1 \mu_1 \left(2\frac{\partial v_1}{\partial y} - \frac{2}{3} \operatorname{div} \overline{V}_1\right) + \frac{\partial}{\partial z} f_1 \mu_1 \left(\frac{\partial w_1}{\partial y} + \frac{\partial v_1}{\partial z}\right) +$$

$$+ \frac{\partial}{\partial x} f_1 \mu_1 \left(\frac{\partial u_1}{\partial y} + \frac{\partial v_1}{\partial x}\right) + K_{12}(v_2 - v_1) + K_{13}(v_3 - v_1) + Y_1 \rho_1$$

$$(5.1)$$

$$\rho_1 \frac{dw_1}{dt} = -f_1 \frac{\partial p}{\partial z} + \frac{\partial}{\partial z} f_1 \mu_1 \left(2 \frac{\partial w_1}{\partial z} - \frac{2}{3} \operatorname{div} \overline{V}_1 \right) + \frac{\partial}{\partial x} f_1 \mu_1 \left(\frac{\partial u_1}{\partial z} + \frac{\partial w_1}{\partial x} \right) +$$

$$+ \frac{\partial}{\partial y} f_1 \mu_1 \left(\frac{\partial v_1}{\partial z} + \frac{\partial w_1}{\partial y} \right) + K_{12} (w_2 - w_1) + K_{13} (w_3 - w_1) + Z_1 \rho_1$$

for the first medium;

$$\rho_2 \frac{du_2}{dt} = -f_2 \frac{\partial p}{\partial x} + \frac{\partial}{\partial x} f_2 \mu_2 \left(2 \frac{\partial u_2}{\partial x} - \frac{2}{3} \operatorname{div} \overline{V}_2 \right) + \frac{\partial}{\partial y} f_2 \mu_2 \left(\frac{\partial v_2}{\partial x} + \frac{\partial u_2}{\partial y} \right) +$$

$$+ \frac{\partial}{\partial z} f_2 \mu_2 \left(\frac{\partial w_2}{\partial x} + \frac{\partial u_2}{\partial z} \right) + K_{12} (u_1 - u_2) + K_{23} (u_3 - u_2) + X_2 \rho_2$$

$$\rho_2 \frac{dv_2}{dt} = -f_2 \frac{\partial p}{\partial y} + \frac{\partial}{\partial y} f_2 \mu_2 \left(2 \frac{\partial v_2}{\partial y} - \frac{2}{3} \operatorname{div} \overline{V}_2 \right) + \frac{\partial}{\partial z} f_2 \mu_2 \left(\frac{\partial w_2}{\partial y} + \frac{\partial v_2}{\partial z} \right) +$$

$$+ \frac{\partial}{\partial x} f_2 \mu_2 \left(\frac{\partial u_2}{\partial y} + \frac{\partial v_2}{\partial x} \right) + K_{12} (v_1 - v_2) + K_{23} (v_3 - v_2) + Y_2 \rho_2$$

$$\rho_2 \frac{dw_2}{dt} = -f_2 \frac{\partial p}{\partial z} + \frac{\partial}{\partial z} f_2 \mu_2 \left(2 \frac{\partial w_2}{\partial z} - \frac{2}{3} \operatorname{div} \overline{V}_2 \right) + \frac{\partial}{\partial x} f_2 \mu_2 \left(\frac{\partial u_2}{\partial z} + \frac{\partial w_2}{\partial x} \right) +$$

$$+ \frac{\partial}{\partial y} f_2 \mu_2 \left(\frac{\partial v_2}{\partial z} + \frac{\partial w_2}{\partial y} \right) + K_{12} (w_1 - w_2) + K_{23} (w_2 - w_3) + Z_2 \rho_2$$

$$(5.2)$$

for the second medium, and

$$\rho_3 \frac{du_3}{dt} = -f_3 \frac{\partial p}{\partial x} + \frac{\partial}{\partial x} f_3 \mu_3 \left(2 \frac{\partial u_3}{\partial x} - \frac{2}{3} \operatorname{div} \overline{V}_3 \right) + \frac{\partial}{\partial y} f_3 \mu_3 \left(\frac{\partial v_3}{\partial x} + \frac{\partial u_3}{\partial y} \right) +$$

$$+ \frac{\partial}{\partial z} f_3 \mu_3 \left(\frac{\partial w_3}{\partial x} + \frac{\partial u_3}{\partial z} \right) + K_{13} (u_1 - u_3) + K_{23} (u_2 - u_3) + X_3 \rho_3$$

$$\rho_3 \frac{dv_3}{dt} = -f_3 \frac{\partial p}{\partial y} + \frac{\partial}{\partial y} f_3 \mu_3 \left(2 \frac{\partial v_3}{\partial y} - \frac{2}{3} \operatorname{div} \overline{V}_3 \right) + \frac{\partial}{\partial z} f_3 \mu_3 \left(\frac{\partial w_3}{\partial y} + \frac{\partial v_3}{\partial z} \right)$$

$$+ \frac{\partial}{\partial z} f_3 \mu_3 \left(\frac{\partial u_3}{\partial y} + \frac{\partial v_3}{\partial x} \right) + K_{13} (v_1 - v_3) + K_{23} (v_2 - v_3) + Y_3 \rho_3$$

$$(5.3)$$

$$\rho_3 \frac{dw_3}{dt} = -f_3 \frac{\partial p}{\partial z} + \frac{\partial}{\partial z} f_3 \mu_3 \left(2 \frac{\partial w_3}{\partial z} - \frac{2}{3} \operatorname{div} \overline{V}_3 \right) + \frac{\partial}{\partial x} f_3 \mu_3 \left(\frac{\partial u_3}{\partial z} + \frac{\partial w_3}{\partial x} \right)$$

$$+ \frac{\partial}{\partial y} f_3 \mu_3 \left(\frac{\partial v_3}{\partial z} + \frac{\partial w_3}{\partial y} \right) + K_{13} (w_1 - w_3) + K_{23} (w_2 - w_3) + Z_3 \rho_3$$

for the third medium, where

$$K_{12} = K_{21};$$
$$K_{13} = K_{31};$$
$$K_{23} = K_{32}.$$

§ 6. Equations of Motion for Viscous Two-Phase Media

For the case of a two-phase flow in a three-dimensional Cartesian coordinate system, the equations of § 5 become

$$\rho_1 \frac{du_1}{dt} = -f_1 \frac{\partial p}{\partial x} + \frac{\partial}{\partial x} f_1 \mu_1 \left(2 \frac{\partial u_1}{\partial x} - \frac{2}{3} \operatorname{div} \bar{V}_1\right) + \frac{\partial}{\partial y} f_1 \mu_1 \left(\frac{\partial v_1}{\partial x} + \frac{\partial u_1}{\partial y}\right) +$$

$$+ \frac{\partial}{\partial z} f_1 \mu_1 \left(\frac{\partial w_1}{\partial x} + \frac{\partial u_1}{\partial z}\right) + K(u_2 - u_1) + X_1 \rho_1$$

$$\rho_1 \frac{dv_1}{dt} = -f_1 \frac{\partial p}{\partial y} + \frac{\partial}{\partial y} f_1 \mu_1 \left(2 \frac{\partial v_1}{\partial y} - \frac{2}{3} \operatorname{div} \bar{V}_1\right) + \frac{\partial}{\partial z} f_1 \mu_1 \left(\frac{\partial w_1}{\partial y} + \frac{\partial v_1}{\partial z}\right) +$$

$$+ \frac{\partial}{\partial x} f_1 \mu_1 \left(\frac{\partial u_1}{\partial y} + \frac{\partial v_1}{\partial x}\right) + K(v_2 - v_1) + Y_1 \rho_1$$

$$\rho_1 \frac{dw_1}{dt} = -f_1 \frac{\partial p}{\partial z} + \frac{\partial}{\partial z} f_1 \mu_1 \left(2 \frac{\partial w_1}{\partial z} - \frac{2}{3} \operatorname{div} \bar{V}_1\right) +$$

$$+ \frac{\partial}{\partial x} f_1 \mu_1 \left(\frac{\partial u_1}{\partial z} + \frac{\partial w_1}{\partial x}\right) + \frac{\partial}{\partial y} f_1 \mu_1 \left(\frac{\partial v_1}{\partial z} + \frac{\partial w_1}{\partial y}\right) + K(w_2 - w_1) + Z_1 \rho_1$$

$$\rho_2 \frac{du_2}{dt} = -f_2 \frac{\partial p}{\partial x} + \frac{\partial}{\partial x} f_2 \mu_2 \left(2 \frac{\partial u_2}{\partial x} - \frac{2}{3} \operatorname{div} \bar{V}_2\right) + \frac{\partial}{\partial y} f_2 \mu_2 \left(\frac{\partial v_2}{\partial x} + \frac{\partial u_2}{\partial y}\right) +$$

$$+ \frac{\partial}{\partial z} f_2 \mu_2 \left(\frac{\partial w_2}{\partial x} + \frac{\partial u_2}{\partial z}\right) + K(u_1 - u_2) + X_2 \rho_2$$

$$\rho_2 \frac{dv_2}{dt} = -f_2 \frac{\partial p}{\partial y} + \frac{\partial}{\partial y} f_2 \mu_2 \left(2 \frac{\partial v_2}{\partial y} - \frac{2}{3} \operatorname{div} \bar{V}_2\right) + \frac{\partial}{\partial z} f_2 \mu_2 \left(\frac{\partial w_2}{\partial y} + \frac{\partial v_2}{\partial z}\right) +$$

$$+ \frac{\partial}{\partial x} f_2 \mu_2 \left(\frac{\partial u_2}{\partial y} + \frac{\partial v_2}{\partial x}\right) + K(v_1 - v_2) + Y_2 \rho_2$$

$$\rho_2 \frac{dw_2}{dt} = -f_1 \frac{\partial p}{\partial z} + \frac{\partial}{\partial z} f_2 \mu_2 \left(2 \frac{\partial w_2}{\partial z} - \frac{2}{3} \operatorname{div} \bar{V}_2\right) + \frac{\partial}{\partial x} f_2 \mu_2 \left(\frac{\partial u_2}{\partial z} + \frac{\partial w_2}{\partial x}\right) +$$

$$+ \frac{\partial}{\partial y} f_2 \mu_2 \left(\frac{\partial v_2}{\partial z} + \frac{\partial w_2}{\partial y}\right) + K(w_1 - w_2) + Z_2 \rho_2$$

$$(6.1)$$

where

$$K_{12} = K_{21} = K.$$

For a two-dimensional, two-phase flow the system (6.1) is written [2]

$$\left.\begin{aligned}
\rho_1 \frac{du_1}{dt} &= -f_1\frac{\partial p}{\partial x} + \frac{\partial}{\partial x}f_1\mu_1\left(2\frac{\partial u_1}{\partial x} - \frac{2}{3}\operatorname{div}\overline{V}_1\right) + \frac{\partial}{\partial y}f_1\mu_1\left(\frac{\partial v_1}{\partial x} + \frac{\partial u_1}{\partial y}\right) + K(u_2 - u_1) + X_1\rho_1 \\[2mm]
\rho_1 \frac{dv_1}{dt} &= -f_1\frac{\partial p}{\partial y} + \frac{\partial}{\partial y}f_1\mu_1\left(2\frac{\partial v_1}{\partial y} - \frac{2}{3}\operatorname{div}\overline{V}_1\right) + \frac{\partial}{\partial x}f_1\mu_1\left(\frac{\partial u_1}{\partial y} + \frac{\partial v_1}{\partial x}\right) + K(v_2 - v_1) + Y_1\rho_1
\end{aligned}\right\} \quad (6.2)$$

for the first medium and

$$\left.\begin{aligned}
\rho_2 \frac{\partial u_2}{\partial t} &= -f_2\frac{\partial p}{\partial x} + \frac{\partial}{\partial x}f_2\mu_2\left(2\frac{\partial u_2}{\partial x} - \frac{2}{3}\operatorname{div}\overline{V}_1\right) + \frac{\partial}{\partial y}f_2\mu_2\left(\frac{\partial v_2}{\partial x} + \frac{\partial u_2}{\partial y}\right) + K(u_1 - u_2) + X_2\rho_2 \\[2mm]
\rho_2 \frac{dv_2}{dt} &= -f_2\frac{\partial p}{\partial y} + \frac{\partial}{\partial y}f_2\mu_2\left(2\frac{\partial v_2}{\partial y} - \frac{2}{3}\operatorname{div}\overline{V}_2\right) + \frac{\partial}{\partial x}f_2\mu_2\left(\frac{\partial u_2}{\partial y} + \frac{\partial v_2}{\partial x}\right) + K(v_1 - v_2) + Y_2\rho_2
\end{aligned}\right\} \quad (6.3)$$

for the second medium.

Let us note that the pressure p in the equations in this section is assumed to be identical for all the phases. However, cases of motion of multiphase media exist in which it is necessary to take account of different pressures for different phases. In order to close such a system of equations, equations connecting these pressures must be added.

§ 7. Equations of Motion of Ideal Multiphase Media

If the coefficients of viscosity in the equations of the preceding section are set equal to zero, the equations of motion of ideal multiphase media are obtained.

The equations of motion of the n-th phase taking part in the motion of an N-phase mixture are

$$\left.\begin{aligned}
\rho_n \frac{du_n}{dt} &= -f_n\frac{\partial p}{\partial x} + \sum_{j=1}^{N}K_{jn}(u_j - u_n) + X_n\,\rho_n \\[2mm]
\rho_n \frac{dv_n}{dt} &= -f_n\frac{\partial p}{\partial y} + \sum_{j=1}^{N}K_{jn}(v_j - v_n) + Y_n\,\rho_n \\[2mm]
\rho_n \frac{dw_n}{dt} &= -f_n\frac{\partial p}{\partial z} + \sum_{j=1}^{N}K_{jn}(w_j - w_n) + Z_n\,\rho_n
\end{aligned}\right\} \quad (7.1)$$

in Cartesian space coordinates.

For incompressible media, i. e., for

$$\rho_{nl} = \text{const},$$

the system of equations (7.1), (3.2), (2.1) is closed.

If the media participating in the mixture motion are compressible, then it is necessary to add

$$p = p(\rho_{ni}, c_n) \quad (n = \overline{1, N}), \tag{7.2}$$

where c_n is a constant associated with the entropy of the medium, to the system for the case of barotropic media.

Since we shall use the equations of motion of ideal three-and two-phase media in succeeding chapters, let us write them for a two-dimensional flow.

The t h r e e - p h a s e flow equations are

$$\left.\begin{aligned}
\rho_1 \frac{du_1}{dt} &= -f_1\frac{\partial p}{\partial x} + K_{12}(u_2 - u_1) + K_{13}(u_3 - u_1) + X_1\rho_1 \\[2mm]
\rho_1 \frac{dv_1}{dt} &= -f_1\frac{\partial p}{\partial y} + K_{12}(v_2 - v_1) + K_{13}(v_3 - v_1) + Y_1\rho_1
\end{aligned}\right\} \tag{7.3}$$

for the first medium,

$$\left.\begin{aligned}
\rho_2 \frac{du_2}{dt} &= -f_2\frac{\partial p}{\partial x} + K_{12}(u_1 - u_2) + K_{23}(u_3 - u_2) + X_2\rho_2 \\[2mm]
\rho_2 \frac{dv_2}{dt} &= -f_2\frac{\partial p}{\partial y} + K_{12}(v_1 - v_2) + K_{23}(v_3 - v_2) + Y_2\rho_2
\end{aligned}\right\} \tag{7.4}$$

for the second medium, and

$$\left.\begin{aligned}
\rho_3 \frac{du_3}{dt} &= -f_3\frac{\partial p}{\partial x} + K_{13}(u_1 - u_3) + K_{23}(u_2 - u_3) + X_3\rho_3 \\[2mm]
\rho_3 \frac{dv_3}{dt} &= -f_3\frac{\partial p}{\partial y} + K_{13}(v_1 - v_3) + K_{23}(v_2 - v_3) + Y_3\rho_3
\end{aligned}\right\} \tag{7.5}$$

for the third medium.

The t w o - p h a s e flow equations are

$$\left.\begin{aligned}
\rho_1 \frac{du_1}{dt} &= -f_1\frac{\partial p}{\partial x} + K(u_2 - u_1) + X_1\rho_1 \\[2mm]
\rho_1 \frac{dv_1}{dt} &= -f_1\frac{\partial p}{\partial y} + K(v_2 - v_1) + Y_1\rho_1
\end{aligned}\right\} \tag{7.6}$$

for the first medium and

$$\left.\begin{aligned}
\rho_2 \frac{du_2}{dt} &= -f_2\frac{\partial p}{\partial x} + K(u_1 - u_2) + X_2\rho_2 \\[2mm]
\rho_2 \frac{dv_2}{dt} &= -f_2\frac{\partial p}{\partial y} + K(v_1 - v_2) + Y_2\rho_2
\end{aligned}\right\} \tag{7.7}$$

for the second medium.

In the case of N-phase mixture motion, still another equation connecting the two different pressures and the other motion parameters must be added to the system of equations of an ideal two-phase mixture if one group of media has a different pressure from that of the other group:

$$p_1 = F(p_2, \rho_1, \rho_2 \ldots \rho_N; \rho_{1i}, \rho_{2i} \ldots \rho_{Ni}).$$

Consequently, the system of equations will be closed.

A number of new aerodynamic equations which may be utilized in studying the motion of mixtures consisting of ideal and viscous media can be formulated by methods analogous to the above. These will be considered in more detail in subsequent chapters in the solution of specific problems.

§ 8. Equations of Motion in Curvilinear Orthogonal Coordinates

Starting from [1], Kleiman [3] derived the equations of motion in curvilinear coordinates. Utilizing orthogonal curvilinear coordinates q_1, q_2, q_3, he isolated a volume element with sides ds_1, ds_2, ds_3 in the mixture, and just as was shown in § 4, composed the total force acting on the n-th phase in the isolated mixture volume in the form

$$\frac{\partial \left(\bar{p}_{n1} ds_2, ds_3 \right)}{\partial q_1} dq_1 + \frac{\partial \left(\bar{p}_{n2} ds_1 ds_2 \right)}{\partial q_2} dq_2 + \frac{\partial \left(\bar{p}_{n3} ds_1 ds_2 \right)}{\partial q_3} dq_3 + \left(\bar{P}'_n + \bar{P}'_n \right) ds_1 ds_2 ds_3 + \bar{F}_n \, \rho_n ds_1 ds_2 ds_3,$$

where $\bar{p}_{n1}$ is a vector equal and opposite to the absolute value of the stress vector on an area perpendicular to the tangent to the coordinate line q_1. The $\bar{p}_{n2}$ and $\bar{p}_{n3}$ are expressed analogously relative to q_2 and q_3. Replacing ds_1, ds_2, ds_3 by the Lamé parameters H_1, H_2, H_3, we obtain the equations of motion of multiphase media

$$\frac{d\bar{v}_n}{dt} = \bar{F}_n + \frac{1}{\rho_n H_1 H_2 H_3} \left[\frac{\partial \left(\bar{p}_{n1} H_2 H_3 \right)}{\partial q_1} + \frac{\partial \left(\bar{p}_{n2} H_1 H_3 \right)}{\partial q_2} + \frac{\partial \left(\bar{p}_{n3} H_1 H_2 \right)}{\partial q_3} \right] + \frac{\bar{P}'_n + \bar{P}'_n}{\rho_n} \tag{8.1}$$

$$(n = 1, \cdots N).$$

This vector equation may be written in projection on the tangents to the coordinate lines. This allows equations of motion to be obtained in Cartesian, cylindrical, and spherical coordinates. In rectangular Cartesian coordinates ($q_1 = x$, $q_2 = y$, $q_3 = z$), (8.1) yields the system (4.4).

Let us write down the equations of motion of viscous multiphase media in cylindrical and spherical coordinates. In cylindrical coordinates ($q_1 = r$, $q_2 = \varphi$, $q_3 = z$)

$$\frac{\partial v_{nr}}{\partial t} + v_{nr} \frac{\partial v_{nr}}{\partial r} + \frac{v_{n\varphi}}{r} \frac{\partial v_{nr}}{\partial \varphi} + v_{nz} \frac{\partial v_{nr}}{\partial z} - \frac{v_{n\varphi}^2}{r} = F_{nr} + \frac{1}{\rho_n} \left(\frac{\partial p_{nrr}}{\partial r} + \frac{1}{r} \frac{\partial p_{n\varphi r}}{\partial \varphi} + \frac{\partial p_{nzr}}{\partial z} + \right.$$

$$\left. + \frac{p_{nrr} - p_{n\varphi\varphi}}{r} \right) + \frac{1}{\rho_n} \sum_{j=1}^{N} K_{jn} (v_{jr} - v_{nr}) + \frac{p}{\rho_n} \frac{\partial f_n}{\partial r}$$

$$\frac{\partial v_{n\varphi}}{\partial t} + v_{nr} \frac{\partial v_{n\varphi}}{\partial r} + \frac{v_{n\varphi}}{r} \frac{\partial v_{n\varphi}}{\partial \varphi} + v_{nz} \frac{\partial v_{n\varphi}}{\partial z} + \frac{v_{nr} v_{n\varphi}}{r} = F_{n\varphi} + \frac{1}{\rho_n} \left(\frac{\partial p_{nr\varphi}}{\partial r} + \frac{1}{r} \frac{\partial p_{n\varphi\varphi}}{\partial \varphi} + \right.$$

$$\left. + \frac{\partial p_{nz\varphi}}{\partial z} + \frac{p_{nr\varphi}}{r} \right) + \frac{1}{\rho_n} \sum_{j=1}^{N} K_{jn} (v_{j\varphi} - v_{n\varphi}) + \frac{p}{\rho_n} \frac{\partial f_n}{r \partial \varphi} \tag{8.2}$$

$$\frac{\partial v_{nz}}{\partial t} + v_{nr} \frac{\partial v_{nz}}{\partial r} + \frac{v_{n\varphi}}{r} \frac{\partial v_{nz}}{\partial \varphi} + v_{nz} \frac{\partial v_{nz}}{\partial z} = F_{nz} +$$

$$+ \frac{1}{\rho_n} \left(\frac{\partial p_{nrz}}{\partial r} + \frac{1}{r} \frac{\partial p_{n\varphi z}}{\partial \varphi} + \frac{\partial p_{nzz}}{\partial z} + \frac{p_{nrz}}{r} \right) + \frac{1}{\rho_n} \sum_{j=1}^{N} K_{jn} (v_{jz} - v_{nz}) + \frac{p}{\rho_n} \frac{\partial f_n}{\partial z}$$

Since

$$p_{nrr} = \left(-p + 2\mu_n \frac{\partial v_{nr}}{\partial r}\right) f_n,$$

$$p_{n\varphi\varphi} = \left[-p + 2\mu_n \left(\frac{1}{r}\frac{\partial v_{n\varphi}}{\partial \varphi} + \frac{v_{nr}}{r}\right)\right] f_n,$$

$$p_{nzz} = \left(-p + 2\mu_n \frac{\partial v_{nz}}{\partial z}\right) f_n,$$

$$p_{n\varphi r} = p_{nr\varphi} = f_n \mu_n \left(\frac{1}{r}\frac{\partial v_{nr}}{\partial \varphi} + \frac{\partial v_{n\varphi}}{\partial r} - \frac{v_{n\varphi}}{r}\right),$$

$$p_{nz\varphi} = p_{n\varphi z} = f_n \mu_n \left(\frac{\partial v_{n\varphi}}{\partial z} + \frac{1}{r}\frac{\partial v_{nz}}{\partial \varphi}\right),$$

$$p_{nrz} = p_{nzr} = f_n \mu_n \left(\frac{\partial v_{nz}}{\partial r} + \frac{\partial v_{nr}}{\partial z}\right),$$

(8.2) may be rewritten as

$$\rho_n \left(\frac{\partial v_{nr}}{\partial t} + v_{nr}\frac{\partial v_{nr}}{\partial r} + \frac{v_{n\varphi}}{r}\frac{\partial v_{nr}}{\partial \varphi} + v_{nz}\frac{\partial v_{nr}}{\partial z} - \frac{v_{n\varphi}^2}{r}\right) = -\frac{\rho_n}{\rho_{ni}}\frac{\partial p}{\partial r} + \frac{\partial}{\partial r}\left\{\frac{\rho_n}{\rho_{ni}}\left(-\frac{2}{3}\mu_n\Theta_n + 2\mu_n\frac{\partial v_{nr}}{\partial r}\right)\right\} +$$

$$+ \frac{1}{r}\frac{\partial}{\partial \varphi}\left[\frac{\rho_n}{\rho_{ni}}\mu_n\left(\frac{1}{r}\frac{\partial v_{nr}}{\partial \varphi} + \frac{\partial v_{n\varphi}}{\partial r} - \frac{v_{n\varphi}}{r}\right)\right] + \frac{\partial}{\partial z}\left[\frac{\rho_n}{\rho_{ni}}\mu_n\left(\frac{\partial v_{nz}}{\partial r} + \frac{\partial v_{nr}}{\partial z}\right)\right] +$$

$$+ \frac{2\mu_n}{r}\frac{\rho_n}{\rho_{ni}}\left(\frac{\partial v_{nr}}{\partial r} - \frac{1}{r}\frac{\partial v_{n\varphi}}{\partial \varphi} - \frac{v_{nr}}{r}\right) + \sum_{j=1}^{N} K_{jn}(v_{mr} - v_{nr}) + \rho_n F_{nr}$$

$$\rho_n \left(\frac{\partial v_{n\varphi}}{\partial t} + v_{nr}\frac{\partial v_{n\varphi}}{\partial r} + \frac{v_{n\varphi}}{r}\frac{\partial v_{n\varphi}}{\partial \varphi} + v_{nz}\frac{\partial v_{n\varphi}}{\partial z} + \frac{v_{n\varphi}v_{nz}}{r}\right) = -\frac{\rho_n}{\rho_{ni}r}\frac{\partial p}{\partial \varphi} +$$

$$+ \frac{1}{r}\frac{\partial}{\partial \varphi}\left\{\frac{\rho_n}{\rho_{ni}}\left[-\frac{2}{3}\mu_n\Theta_n + 2\mu_n\left(\frac{1}{r}\frac{\partial v_{n\varphi}}{\partial \varphi} + \frac{v_{nr}}{r}\right)\right]\right\} + \frac{\partial}{\partial r}\left[\frac{\rho_n}{\rho_{ni}}\mu_n\left(\frac{1}{r}\frac{\partial v_{nr}}{\partial \varphi} + \frac{\partial v_{n\varphi}}{\partial r} - \frac{v_{n\varphi}}{r}\right)\right] +$$

$$+ \frac{\partial}{\partial z}\left[\frac{\rho_n}{\rho_{ni}}\mu_n\left(\frac{\partial v_{n\varphi}}{\partial z} + \frac{1}{r}\frac{\partial v_{nz}}{\partial \varphi}\right)\right] + \frac{2\mu_n}{r}\frac{\rho_n}{\rho_{ni}}\left(\frac{1}{r}\frac{\partial v_{nr}}{\partial \varphi} + \frac{\partial v_{n\varphi}}{\partial r} - \frac{v_{n\varphi}}{r}\right) + \sum_{j=1}^{N} K_{jn}(v_{m\varphi} - v_{n\varphi}) + \rho_n F_{n\varphi}$$

(8.3)

$$\rho_n \left(\frac{\partial v_{nz}}{\partial t} + v_{nr}\frac{\partial v_{nz}}{\partial r} + \frac{v_{n\varphi}}{r}\frac{\partial v_{nz}}{\partial \varphi} + v_{nz}\frac{\partial v_{nz}}{\partial z}\right) = -\frac{\rho_n}{\rho_{ni}}\frac{\partial p}{\partial z} + \frac{\partial}{\partial z}\left\{\frac{\rho_n}{\rho_{ni}}\left[-\frac{2}{3}\mu_n\Theta_n + 2\mu_n\frac{\partial v_{nz}}{\partial z}\right]\right\} +$$

$$+ \frac{\partial}{\partial r}\left[\frac{\rho_n}{\rho_{ni}}\mu_n\left(\frac{\partial v_{nz}}{\partial r} + \frac{\partial v_{nr}}{\partial z}\right)\right] + \frac{1}{r}\frac{\partial}{\partial \varphi}\left[\frac{\rho_n}{\rho_{ni}}\mu_n\left(\frac{\partial v_{n\varphi}}{\partial z} + \frac{1}{r}\frac{\partial v_{nz}}{\partial \varphi}\right)\right] +$$

$$+ \frac{\rho_n}{\rho_{ni}}\frac{\mu_n}{r}\left(\frac{\partial v_{nz}}{\partial r} + \frac{\partial v_{nr}}{\partial z}\right) + \sum_{j=1}^{N} K_{jn}(v_{mz} - v_{nz}) + \rho_n F_{nz}$$

$$\Theta = \operatorname{div} \overline{V}$$

In spherical coordinates ($q_1 = r$, $q_2 = \varphi$, $q_3 = \psi$)

$$\frac{\partial v_{nr}}{\partial t} + v_{nr}\frac{\partial v_{nr}}{\partial r} + \frac{v_{n\varphi}}{r}\frac{\partial v_{nr}}{\partial \varphi} + \frac{v_{n\psi}}{r\sin\varphi}\frac{\partial v_{nr}}{\partial \psi} - \frac{v_{n\varphi}^2 + v_{n\psi}^2}{r} = F_{nr} + \frac{1}{\rho_n}\left[\frac{\partial p_{nrr}}{\partial r} + \frac{1}{r}\frac{\partial p_{n\varphi r}}{\partial \Theta} + \right.$$

$$+ \frac{1}{r\sin\varphi}\cdot\frac{\partial p_{n\psi r}}{\partial \psi} + \frac{1}{r}\left(2p_{nrr} + p_{n\varphi r}\cot\varphi - p_{n\varphi\varphi} - p_{n\psi\psi}\right)\bigg] + \frac{1}{\rho_n}\sum_{j=1}^{N} K_{jn}(v_{jr} - v_{nr}) + \frac{p}{\rho_n}\frac{\partial f_n}{\partial r}$$

$$\frac{\partial v_{n\varphi}}{\partial t} + v_{nr}\frac{\partial v_{n\varphi}}{\partial r} + \frac{v_{n\varphi}}{r}\frac{\partial v_{n\varphi}}{\partial \varphi} + \frac{v_{n\psi}}{r\sin\varphi}\frac{\partial v_{n\varphi}}{\partial \psi} + \frac{v_{nr}-v_{n\varphi}}{r} - \frac{v_{n\psi}^2}{r}\cot\varphi = F_{n\varphi} + \frac{1}{\rho_n}\left[\frac{\partial p_{nr\varphi}}{\partial r} + \right.$$

$$\left. + \frac{1}{r}\frac{\partial p_{n\varphi\varphi}}{\partial \varphi} + \frac{1}{r\sin\varphi}\frac{\partial p_{n\psi\psi}}{\partial \psi} + \frac{1}{r}\left(p_{n\varphi\varphi}\cot\varphi + 3p_{n\varphi r} - p_{n\psi\psi}\cot\varphi\right)\right] + \frac{1}{\rho_n}\sum_{j=1}^{N}K_{jn}(v_{j\varphi}-v_{n\varphi}) + \frac{p}{\rho_n}\frac{\partial f_n}{r\partial\varphi}$$

$$\frac{\partial v_{n\psi}}{\partial t} + v_{nr}\frac{\partial v_{n\psi}}{\partial r} + \frac{v_{n\varphi}}{r}\frac{\partial v_{n\psi}}{\partial \theta} + \frac{v_{n\psi}}{r\sin\varphi}\frac{\partial v_{n\psi}}{\partial \psi} + \frac{v_{n\psi}v_{nr}}{r} + \frac{v_{n\psi}v_{n\varphi}}{r}\cot\varphi = F_{n\psi} + \frac{1}{\rho_n}\times$$

$$\times\left[\frac{\partial p_{nr\psi}}{\partial r} + \frac{1}{r}\frac{\partial p_{n\varphi\psi}}{\partial \varphi} + \frac{1}{r\sin\varphi}\frac{\partial p_{n\psi\psi}}{\partial \psi} + \frac{1}{r}\left(2p_{n\varphi\psi}\cot\varphi + 3p_{n\psi r}\right)\right] + \frac{1}{\rho_n}\sum_{j=1}^{N}K_{jn}(v_{j\varphi}-v_{n\psi}) + \frac{p}{\rho_n}\frac{\partial f_n}{r\sin\varphi\,\partial\psi}$$

$$(n = 1,\ldots,N)$$

$$\left.\right\} \quad (8.4)$$

where

$$p_{nrr} = f_n\left(-p + 2\mu_n\frac{\partial v_{nr}}{\partial r}\right),$$

$$p_{n\varphi\varphi} = f_n\left[-p + 2\mu_n\left(\frac{1}{r}\frac{\partial v_{n\varphi}}{\partial \varphi} + \frac{v_{nr}}{r}\right)\right],$$

$$p_{n\psi\psi} = f_n\left[-p + 2\mu_n\left(\frac{1}{r\sin\varphi}\frac{\partial v_{n\psi}}{\partial \psi} + \frac{v_{nr}}{r} + \frac{v_{n\varphi}\cot\varphi}{r}\right)\right],$$

$$p_{n\varphi\psi} = p_{n\psi\varphi} = f_n\mu_n\left(\frac{1}{r\sin\varphi}\frac{\partial v_{n\varphi}}{\partial \psi} + \frac{1}{r}\frac{\partial v_{n\psi}}{\partial \varphi} - \frac{v_{n\psi}\cot\varphi}{r}\right),$$

$$p_{n\psi r} = p_{nr\psi} = f_n\mu_n\left(\frac{\partial v_{n\psi}}{\partial r} + \frac{1}{r\sin\varphi}\frac{\partial v_{nr}}{\partial \psi} - \frac{v_{n\psi}}{r}\right),$$

$$p_{nr\varphi} = p_{n\varphi r} = f_n\mu_n\left(\frac{1}{r}\frac{\partial v_{nr}}{\partial \varphi} + \frac{\partial v_{n\varphi}}{\partial r} - \frac{v_{n\varphi}}{r}\right).$$

§ 9. Equations of Motion of Ideal Multiphase Media in Cylindrical and Spherical Coordinates

The equations of motion of viscous–ideal and ideal multiphase media are easily obtained by utilizing the equations of motion (8.3), (8.4) of viscous media; the ideal phases will here have zero coefficients of viscosity. Then the equations of motion of the ideal multiphase media will be

$$\frac{\partial v_{nr}}{\partial t} + v_{nr}\frac{\partial v_{nr}}{\partial r} + \frac{v_{n\varphi}}{r}\frac{\partial v_{nr}}{\partial \varphi} + v_{nz}\frac{\partial v_{nr}}{\partial z} - \frac{v_{n\varphi}^2}{r} = F_{nr} - \frac{1}{\rho_{ni}}\frac{\partial p}{\partial r} + \frac{1}{\rho_n}\sum_{j=1}^{N}K_{jn}(v_{jr}-v_{nr})$$

$$\frac{\partial v_{n\varphi}}{\partial t} + v_{nr}\frac{\partial v_{n\varphi}}{\partial r} + \frac{v_{n\varphi}}{r}\frac{\partial v_{n\varphi}}{\partial \varphi} + v_{nz}\frac{\partial v_{n\varphi}}{\partial z} + \frac{v_{nr}v_{n\varphi}}{r} = F_{n\varphi} - \frac{1}{\rho_{ni}}\frac{\partial p}{r\partial\varphi} + \frac{1}{\rho_n}\sum_{j=1}^{N}K_{jn}(v_{j\varphi}-v_{n\varphi})$$

$$\frac{\partial v_{nz}}{\partial t} + v_{nr}\frac{\partial v_{nz}}{\partial r} + \frac{v_{n\varphi}}{r}\frac{\partial v_{nz}}{\partial \varphi} + v_{nz}\frac{\partial v_{nz}}{\partial z} = F_{nz} - \frac{1}{\rho_{ni}}\frac{\partial p}{\partial z} + \frac{1}{\rho_n}\sum_{j=1}^{N}K_{jn}(v_{jz}-v_{nz})$$

$$\left.\right\} \quad (9.1)$$

in cylindrical coordinates, and

$$\frac{\partial v_{nr}}{\partial t} + v_{nr}\frac{\partial v_{nr}}{\partial r} + \frac{v_{n\varphi}}{r}\frac{\partial v_{nr}}{\partial \varphi} + \frac{v_{n\psi}}{r\sin\varphi}\frac{\partial v_{nr}}{\partial \psi} - \frac{v_{n\varphi}^2 + v_{n\psi}^2}{r} = F_{nr} - \frac{1}{\rho_{ni}}\frac{\partial p}{\partial r} + \frac{1}{\rho_n}\sum_{j=1}^{N} K_{jn}\left(v_{jr} - v_{nr}\right)$$

$$\frac{\partial v_{n\varphi}}{\partial t} + v_{nr}\frac{\partial v_{n\varphi}}{\partial r} + \frac{v_{n\varphi}}{r}\frac{\partial v_{n\varphi}}{\partial \theta} + \frac{v_{n\psi}}{r\sin\varphi}\frac{\partial v_{n\psi}}{\partial \psi} + \frac{v_{nr}v_{n\varphi}}{r} - \frac{v_{n\psi}^2}{r}\cot\varphi = F_{n\varphi} - \frac{1}{\rho_{ni}}\frac{\partial p}{r\partial \varphi} + \frac{1}{\rho_n}\sum_{j=1}^{N} K_{jn}\left(v_{j\varphi} - v_{n\varphi}\right)$$

$$\frac{\partial v_{n\psi}}{\partial t} + v_{nr}\frac{\partial v_{n\psi}}{\partial r} + \frac{v_{n\varphi}}{r}\frac{\partial v_{n\psi}}{\partial \varphi} + \frac{v_{n\psi}}{r\sin\varphi}\frac{\partial v_{n\psi}}{\partial \psi} + \frac{v_{n\psi}v_{nr}}{r} + \frac{v_{n\psi}v_{n\varphi}}{r}\cot\varphi = F_{n\psi} - \frac{1}{\rho_{n\psi}}\frac{1}{r\sin\varphi}\frac{\partial p}{\partial \psi} + $$

$$+ \frac{1}{\rho_n}\sum_{j=1}^{N} K_{jn}\left(v_{j\psi} - v_{n\psi}\right)$$

$$(n = 1, \ldots, N)$$

$$\left.\right\}\ (9.2)$$

in spherical coordinates.

CHAPTER II

STEADY MOTION OF IDEAL MULTIPHASE MEDIA IN CONDUITS

The problem of the steady motion of a two-phase flow in conduits of variable and constant cross section was first solved in a linearized formulation by Kleiman [4], who studied the motion of compressible and incompressible media in detail. Bakhriev [5] solved this problem for identical initial velocities by the method of reduction to an integrodifferential equation and extended it to the three-phase flow case. Latipov [6] solved the problem of incompressible two-phase motion in a nonlinear formulation. Let us now consider the solution of these problems.

§ 10. Motion of a Two-Phase Flow*

Let there be one-dimensional steady motion of an ideal two-phase fluid in a variable-cross-section conduit whose axis coincides with the 0x axis and makes an angle ε with the vertical. Let us assume that the cross-sectional area $f(x)$ of the conduit changes only slightly. We take account of the gravity component of each phase along the 0x axis, and we neglect it along the 0y axis. Then the equations of motion (7.6), (7.7) and Lomonosov's law in Euler form (3.2) will become

$$
\left.
\begin{aligned}
u_1 \frac{du_1}{dx} + \frac{1}{\rho_{1i}} \frac{dp}{dx} + g\cos\varepsilon - \frac{K}{\rho_1}(u_2 - u_1) &= 0 \\
u_2 \frac{du_2}{dx} + \frac{1}{\rho_{2i}} \frac{dp}{dx} + g\cos\varepsilon - \frac{K}{\rho_2}(u_1 - u_2) &= 0 \\
\rho_1 u_1 f(x) &= c_1 \\
\rho_2 u_2 f(x) &= c_2
\end{aligned}
\right\}
\tag{10.1}
$$

where c_1 and c_2 are constants, and $f(x)$ is the cross section of the conduit. Expression (2.1) and the equations of state in the case of barotropic fluids are

$$
\frac{\rho_1}{\rho_{1i}} + \frac{\rho_2}{\rho_{2i}} = 1,
\tag{10.2}
$$

$$
\left.
\begin{aligned}
p &= F_1(\rho_{1i}, p_0, \rho_{1i}^0) \\
p &= F_2(\rho_{2i}, p_0, \rho_{2i}^0)
\end{aligned}
\right\}
\tag{10.3}
$$

where p_0, ρ_{1i}^0, ρ_{2i}^0 are the initial values of the pressure and the true densities.

In [4] Kleiman assumes that in many cases the changes in the mixture parameters during the motion are small compared with the magnitudes of these parameters; hence he linearizes the equations.

---

*The Kleiman solution is elucidated in this section.

Let

$$p = p_0 + p', \quad u_1 = u_{10} + u_1',$$
$$u_2 = u_{20} + u_2', \quad \rho_1 = \rho_{10} + \rho_1',$$
$$\rho_2 = \rho_{20} + \rho_2', \quad \rho_{1i} = \rho_{1i}^0 + \rho_{1i}',$$
$$\rho_{2i} = \rho_{2i}^0 + \rho_{2i}', \quad f(x) = f_0 + f',$$

where p', u_1', u_2', ρ_1', ρ_2', ρ_{1i}', ρ_{2i}', f' are very small, and vanish in the initial section. After linearization we have

$$\left. \begin{array}{l} u_{10} \dfrac{du_1'}{dx} + \dfrac{1}{\rho_{1i}^0} \dfrac{dp'}{dx} + g \cos \varepsilon - \dfrac{K}{\rho_{10}}(u_{20} - u_{10}) - \dfrac{K}{\rho_{10}}(u_2' - u_1') = 0 \\[2mm] u_{20} \dfrac{du_2'}{dx} + \dfrac{1}{\rho_{2i}^0} \dfrac{dp}{dx} + g \cos \varepsilon - \dfrac{K}{\rho_{20}}(u_{10} - u_{20}) - \dfrac{K}{\rho_{20}}(u_1' - u_2') = 0 \end{array} \right\} \quad (10.4)$$

$$\left. \begin{array}{l} \rho_{10} u_{10} f' + \rho_{10} f_0 u_1' + u_{10} f_0 \rho_1' = 0 \\[2mm] \rho_{20} u_{20} f' + \rho_{20} f_0 u_2' + u_{20} f_0 \rho_2' = 0 \end{array} \right\} \quad (10.5)$$

Moreover

$$\frac{\rho_1'}{\rho_{1i}^0} + \frac{\rho_2'}{\rho_{2i}^0} - \rho_{1i}' \frac{\rho_{10}}{\rho_{1i}^{02}} - \rho_{2i}' \frac{\rho_{20}}{\rho_{2i}^{02}} = 0, \quad (10.6)$$

$$\left. \begin{array}{l} p' = a_{01}^2 \rho_{1i}' \\[2mm] p' = a_{02}^2 \rho_{2i}' \end{array} \right\} \quad (10.7)$$

where

$$a_{01}^2 = F_1'(\rho_{1i}^0, p_0, \rho_{1i}^0);$$
$$a_{02}^2 = F_2'(\rho_{2i}^0, p_0, \rho_{2i}^0).$$

We then have from (10.6)

$$\left. \begin{array}{l} \dfrac{\rho_1'}{\rho_{1i}^0} + \dfrac{\rho_2'}{\rho_{2i}^0} - \xi p' = 0 \\[3mm] \left(\xi = \dfrac{\rho_{10}}{\rho_{1i}^{02} a_{01}^2} + \dfrac{\rho_{20}}{\rho_{2i}^{02} a_{02}^2} \right) \end{array} \right\} \quad (10.8)$$

We multiply the first equation of the system (10.4) by ρ_{10}, the second by ρ_{20}, and add. We integrate the obtained differential equation under the conditions

$$p' = 0, \quad u_1' = u_2' = 0,$$

when $x = 0$. We then have

$$p' = -xg(\rho_{10} + \rho_{20}) \cos \varepsilon - \rho_{10} u_{10} u_1' - \rho_{20} u_{20} u_2'. \quad (10.9)$$

Substituting ρ_1', ρ_2', from (10.5) into (10.8), we have

$$p' = -\frac{1}{\xi}\left(\frac{\rho_{10}}{\rho_{1i}^0 u_{10}}u_1' + \frac{\rho_{20}}{\rho_{2i}^0 u_{20}}u_2' + \frac{f'}{f_0}\right). \tag{10.10}$$

From (10.9) and (10.10) we determine

$$\left.\begin{aligned}
u_1' &= -\frac{\rho_{1i}^0 u_{10}}{\rho_{10}\,(\rho_{1i}^0 u_{10}^2 - \rho_{2i}^0 u_{20}^2)}\left[(1-\xi\rho_{2i}^0 u_{20}^2)p' - \frac{f'}{f_0}\rho_{2i}^0 u_{20}^2 + xg\,(\rho_{10}+\rho_{20})\cos\varepsilon\right]\\
u_2' &= \frac{\rho_{2i}^0 u_{20}}{\rho_{20}\,(\rho_{1i}^0 u_{10}^2 - \rho_{2i}^0 u_{20}^2)}\left[(1-\xi\rho_{1i}^0 u_{10}^2)p' - \frac{f'}{f_0}\rho_{1i}^0 u_{10}^2 + xg\,(\rho_{10}+\rho_{20})\cos\varepsilon\right]
\end{aligned}\right\} \tag{10.11}$$

where u_1', u_2' are expressed in terms of p', whose value we determine as follows. We multiply the first equation in (10.4) by ρ_{1i}^0, and the second by ρ_{2i}^0. Their difference yields a differential equation which, taking account of (10.11), becomes

$$\frac{dp'}{dx} + K\beta p' = R(x), \tag{10.12}$$

where

$$R(x) = \frac{1}{A}\left\{\frac{df'}{f_1 dx} + K\left(\frac{1}{\rho_{10}u_{10}} + \frac{1}{\rho_{20}u_{20}}\right)\frac{f'}{f_0} + K(u_{10}-u_{20})\left(\frac{1}{\rho_{2i}^0 u_{20}^2} - \frac{1}{\rho_{1i}^0 u_{10}^2}\right) - \right.$$
$$\left. - g\left[M^2 + \frac{\rho_{10}\rho_{20}\,(u_{10}-u_{20})^2}{\rho_{1i}^0\rho_{2i}^0 u_{10}^2 u_{20}^2}\right]\cos\varepsilon - \frac{KgM\cos\varepsilon}{u_{10}u_{20}}\left(\frac{1}{\rho_{10}} + \frac{1}{\rho_{20}}\right)x\right\};$$

$$M = \frac{\rho_{10}}{\rho_{1i}^0 u_{10}} + \frac{\rho_{20}}{\rho_{2i}^0 u_{20}};$$

$$\beta = \frac{B}{A};$$

$$A = \frac{\rho_{10}}{\rho_{1i}^{02} u_{10}^2} + \frac{\rho_{20}}{\rho_{2i}^{02} u_{20}^2} - \xi;$$

$$B = \frac{M - \xi\,(\rho_{10}u_{10}+\rho_{20}u_{20})}{\rho_{10}\rho_{20}u_{10}u_{20}}.$$

The solution of the equation under the condition p' = 0 when x = 0 will be

$$p' = e^{-K\beta x}\int_0^x R(x)\,e^{K\beta x}\,dx. \tag{10.13}$$

The equation obtained affords a foundation for the determination of the velocity and the reduced densities.

Let us consider mixture motion in a conical conduit. In this case

$$f' = 2\pi rx\tan\alpha,$$

where $r = \sqrt{f_0/\pi}$ is the radius of the section corresponding to the origin, and a is the angle formed by the generator of the cone and the axis of the conduit. Then

$$\frac{f'}{f_0} = \eta x, \quad \eta = 2\sqrt{\frac{\pi}{f_0}}\tan\alpha.$$

Finding (R(x), we substitute it into (10.13), and then integrating, we obtain

$$p' = \frac{1}{KB}\left[L_1 x - \frac{L_2}{B}\left(\frac{1}{\rho_{2i}^0 u_{20}^2} - \frac{1}{\rho_{1i}^0 u_{20}^2} \right)(1 - e^{-K\beta x}) \right],$$
(10.14)

where

$$L_1 = K\left[\eta\left(\frac{1}{\rho_{10} u_{10}} + \frac{1}{\rho_{20} u_{20}} \right) - \frac{gM\cos\varepsilon}{u_{10} u_{20}}\left(\frac{1}{\rho_{10}} + \frac{1}{\rho_{20}} \right) \right],$$

$$L_2 = \eta\left(\frac{1}{\rho_{2i}^0 u_{20}} - \frac{1}{\rho_{1i}^0 u_{10}} \right) - KB(u_{10} - u_{20}) - \frac{g\cos\varepsilon}{u_{10} u_{20}}\left[\xi(u_{10} - u_{20}) + M\left(\frac{1}{\rho_{2i}^0} - \frac{1}{\rho_{1i}^0} \right) \right].$$

Substituting the value of p' from (10.14) into (10.11), we obtain u_1', u_2':

$$\left. \begin{aligned}
u_1' &= \frac{1}{B\rho_{10} u_{10}}\left\{ \frac{L_2}{KB}\left(\frac{1}{\rho_{2i}^0 u_{20}^2} - \xi \right)(1 - e^{-K\beta x}) - \frac{x}{\rho_{20} u_{20}}[\eta - g\xi(\rho_{10} + \rho_{20})\cos\varepsilon] \right\} \\
u_2' &= \frac{1}{B\rho_{20} u_{20}}\left\{ \frac{L_2}{KB}\left(\frac{1}{\rho_{1i}^0 u_{10}^2} - \xi \right)(1 - e^{-K\beta x}) + \frac{x}{\rho_{10} u_{10}}[\eta - g\xi(\rho_{10} + \rho_{20})\cos\varepsilon] \right\}
\end{aligned} \right\}$$
(10.15)

On the basis of (10.15) we find the formula for the difference in velocities:

$$u_1' - u_2' = \frac{L_2}{KB}(1 - e^{-K\beta x}).$$
(10.16)

Let us now determine the change in the mass composition of the mixture. Since the reduced densities are variable, the mass composition of the mixture characterized by

$$\gamma = \frac{m_1}{m_2} = \frac{\rho_1}{\rho_2}$$

(m_1, m_2 are the masses of the first and second phases in a given mixture volume) will also be variable depending on x. After linearization, taking account of (10.5), this change may be expressed as

$$\gamma' = \frac{\rho_1}{\rho_2} - \frac{\rho_{10}}{\rho_{20}} = -\frac{1}{B\rho_{2i}^0 u_{10}^2 u_{20}^2}\left\{ \frac{L_2}{KB}\left[1 - \xi(\rho_{10} u_{10}^2 + \rho_{20} u_{20}^2) \right](1 - e^{K\beta x}) + \right.$$

$$\left. + (u_{10} - u_{20})[\eta - g\xi(\rho_{10} + \rho_{20})\cos\varepsilon] x \right\}.$$
(10.17)

Formulas (10.15) − (10.17) express the changes in media velocities, as well as in mass composition of the mixture along the length of a variable-section conduit. However, it is impossible to conduct investigations for as large a value of x as desired on the basis of these formulas [4], if only because f' grows with the increase in x even for very small angles α; therefore, starting with some value of x, the assumption of very small f' is unrealistic, and the values of the motion parameters (velocities, pressures, etc.) obtained are true only in definite finite intervals near the initial section. Summarizing, it may be asserted that it is impossible in investigating the motion parameters in variable-section conduits to limit oneself to a linear formulation which yields the solution near the initial section.

An analysis of mixture motion along cylindrical horizontal conduits is of great theoretical and practical interest.

<u>Compressible Medium</u>. Since $\eta = 0$, $\cos \varepsilon = 0$, (10.14) and (10.15) become

$$p' = -\frac{\rho_{10}\,\rho_{20}\,(u_{10}-u_{20})\,\left(\rho_{1i}^{0}\,u_{10}^{2} - \rho_{2i}^{0}\,u_{20}^{2}\right)}{\rho_{1i}^{0}\,\rho_{2i}^{0}\,u_{10}\,u_{20}\,[M-\xi\,(\rho_{10}\,u_{10} - \rho_{20}\,u_{20})]}\,(1-e^{-K\beta x}), \tag{10.18}$$

$$\left. \begin{aligned} u_{1}' &= -\frac{\rho_{20}\,(u_{10}-u_{20})\,(1-\xi\,\rho_{2i}\,u_{20}^{2})}{\rho_{2i}\,u_{20}\,[M-\xi\,(\rho_{10}\,u_{10}-\rho_{20}\,u_{20})]}\,(1-e^{-K\beta x}) \\[2mm] u_{2}' &= \frac{\rho_{10}\,(u_{10}-u_{20})\,(1-\xi\,\rho_{1i}\,u_{10}^{2})}{\rho_{1i}\,u_{10}\,[M-\xi\,(\rho_{10}\,u_{10}-\rho_{20}\,u_{20})]}\,(1-e^{-K\beta x}) \end{aligned} \right\}. \tag{10.19}$$

From (10.19) we have

$$u_{1} - u_{2} = (u_{10} - u_{20})\,e^{-K\beta x}. \tag{10.20}$$

It is hence seen that the velocity and pressure tend to constant values as $x \to \infty$, where the difference in velocities tends to zero.

Kleiman [4] makes a number of valuable deductions.

1. If the initial values of the velocity heads of both phases are greater or less than $1/\xi$, the lesser velocity increases, and the greater decreases (Fig. 1); if one of the quantities $\rho_{1i}^{0}\,u_{10}^{2}$ and $\rho_{2i}^{0}\,u_{20}^{2}$ is greater and the other less then $1/\xi$, then both velocities increase or decrease. For $u_{10} > u_{20}$ the above may be represented as shown in Fig. 2.

Fig. 1   Fig. 2

In the case considered the change in motion parameters is due to the interaction of the media. If the initial velocities are identical, this interaction is missing, and all the parameters remain constant.

2. The pressure along the conduit does not change, and when the initial values of the velocity heads of the phases are identical, the velocities then vary in this case. If the initial velocity heads are selected so that each equals $1/\xi$, the velocities of the media and the pressure remain constant.

As is seen, changes in the motion parameters up to infinite limits can be determined for mixture motion in horizontal cylindrical conduits by using the linearized equations. However, even in this case there are phenomena which require investigation in a nonlinear formulation. For example, let us consider the case for which the velocity head is $1/2\xi$ for the first medium in the initial section, but not for the second. Then the velocity of the second medium should remain constant although the velocity of the first medium changes. This is actually impossible since the velocity of the second medium would also change as a result of phase interaction.

Let us determine the length of a conduit in which phase mixing could be considered practically complete. In this case (10.17) becomes

$$\gamma' = \frac{u_{10}-u_{20}}{B\,\rho_{20}^{2}\,u_{10}^{2}\,u_{20}^{2}}\,\left[1-\xi\,(\rho_{10}\,u_{10}^{2} + \rho_{20}\,u_{20}^{2})\right]\,(1-e^{-K\beta x}).$$

We can write this as

$$\gamma' = \gamma'_{\infty}(1 - e^{-K\beta x}), \tag{10.21}$$

where γ'_{∞} is the value of γ' as $x \to \infty$:

$$\gamma'_{\infty} = \frac{u_{10}\, u_{20}}{B\rho_{20}^2\, u_{10}^2\, u_{20}^2} \left[1 - \xi\, (\rho_{10}\, u_{10}^2 + \rho_{20}\, u_{20}^2) \right].$$

Let us determine the coordinate x_{∞} of the section in which the quantity γ' differs sufficiently slightly from the value of γ'_{∞}, i.e., is the given part of γ'_{∞}:

$$\gamma' = \chi\, \gamma'_{\infty}.$$

Comparing this with (10.21), we have

$$\chi = 1 - e^{-K\beta x},$$

hence

$$x_{\infty} = -\frac{1}{K\beta} \ln (1 - \chi),$$

where χ is a quantity close to 1.

Let us consider the case when the mixture is made up of incompressible media. We will then have for a horizontal cylindrical conduit

$$\eta = 0, \quad \xi = 0, \quad \cos \varepsilon = 0.$$

From (10.18) and (10.19) we obtain

$$p' = \frac{\rho_{10}\, \rho_{20}\, (u_{10} - u_{20})\, (\rho_{1i}\, u_{10}^2 - \rho_{1i}\, u_{20}^2)}{\rho_{1i}\, \rho_{2i}\, u_{10}\, u_{20}\, M} (1 - e^{-K\beta x}),$$

$$u_1' = -\frac{\rho_{10}\, (u_{10} - u_{20})}{\rho_{2i}\, u_{20}\, M} (1 - e^{-K\beta x}),$$

$$u_2' = \frac{\rho_{10}\, (u_{10} - u_{20})}{\rho_{1i}\, u_{10}\, M} (1 - e^{-K\beta x}).$$

On the basis of the last two equations we have

$$\frac{u_1'}{u_2'} = -\frac{\rho_{20}\, \rho_{1i}}{\rho_{10}\, \rho_{2i}} \frac{u_{10}}{u_{20}},$$

i. e., changes in the phase velocities are directly proportional to the initial values of the volume content of the phases, where, if the velocity of one phase increases, the velocity of the other decreases.

If $x \to \infty$, the velocities of both phases tend to a constant and their difference to zero.

We have characterized the motion in horizontal cylindrical conduits, where the influence of the gravity of the media has been neglected.

Let us note that the determination of the length of conduit where the velocity of the heavy phase reaches a zero value (necessary for the analysis of a gas air lift) is of enormous practical value in extending the problems elucidated above to an ascending flow in an inclined infinite cylindrical conduit.

§ 11. Solution of Problems of Motion of Ideal Incompressible Two-Phase Media in Horizontal Conduits of Variable and Constant Cross Section without Linearization of the Equations*

Let us introduce the notation

$$v_1 = u_1 f(x), \quad v_2 = u_2 f(x)$$

into the system of equations (10.1). Then, subtracting the second equation from the first, we obtain

$$\rho_{1i}\frac{v_1}{f(x)}\frac{dv_1}{dx} - \rho_{2i}\frac{v_2}{f(x)}\frac{dv_2}{dx} - \frac{f'(x)}{f^2(x)}\left(\rho_{1i}v_1^2 - \rho_{2i}v_{2i}^2\right) = \frac{K}{f_1 f_2}(v_2 - v_1), \tag{11.1}$$

where $f'(x)$ is the derivative of $f(x)$ with respect to x. From (10.1) and (10.2) we determine

$$\left.\begin{array}{l} f_1 = \frac{c_1}{\rho_{1i}}\frac{1}{v_1} \\[2mm] v_2 = \frac{c_2}{\rho_{2i}}\frac{\rho_{1i}v_1}{\rho_{1i}v_1 - c_1} \\[2mm] f_2 = \frac{\rho_{1i}v_1 - c_1}{\rho_{1i}v_1} \end{array}\right\} \tag{11.2}$$

Let us consider the motion in a variable-section conduit. Limiting himself to the case of motion for which terms containing $[f'(x)/f^2(x)](\rho_{1i}v_1^2 - \rho_{2i}v_2^2)$ may be neglected in the equations of motion, Latipov obtained from (11.1)

$$\rho_{1i}\frac{v_1}{f(x)}\frac{dv_1}{dx} - \rho_{2i}\frac{v_2}{f(x)}\frac{dv_2}{dx} = \frac{K}{f_1 f_2}(v_2 - v_1). \tag{11.3}$$

Substituting (11.2) into (11.3), we have after some manipulation

$$\frac{\left[\left(v_1 - \frac{c_1}{\rho_{1i}}\right)^3 + \frac{c_1}{\rho_{1i}^2}\frac{c_2^2}{\rho_{2i}}\right]c_1\,dv_1}{K\left(v_1 - \frac{c_1}{\rho_{1i}} - \frac{c_2}{\rho_{2i}}\right)\left(v_1 - \frac{c_1}{\rho_{1i}}\right)v_1^2\,dx} = -f(x). \tag{11.4}$$

Thus, the problem has been reduced to the integration of a first-order ordinary differential equation with separable variables.

Decomposing the rational fraction on the left side of (11.4) into the sum of elementary fractions, we rewrite it as

$$\left[\frac{A}{c_1 v_1} + \frac{B_1}{v_1^2} - \frac{c_2}{c_1\left(v_1 - \frac{c_1}{\rho_{1i}}\right)} + \frac{D}{c_1\left(v_1 - \frac{c_1}{\rho_{1i}} - \frac{c_2}{\rho_{2i}}\right)}\right]\frac{dv_1}{dx} = -\frac{K}{c_1}f(x).$$

---

*The Latipov solution is elucidated in this section.

Let the value of the velocity $u_1 = u_{10}$ be given at the initial section (for $x = x_0$). Then evidently

$$v_1 = v_{10} = v_{10} f(x_0).$$

Integration yields

$$\frac{A}{c_1} \ln \frac{v_1}{v_{10}} - \frac{B_1}{v_1} + \frac{B_1}{v_{10}} - \frac{c_2}{c_1} \ln \frac{v_1 - \dfrac{c_1}{\rho_{1i}}}{v_{10} - \dfrac{c_1}{\rho_{1i}}} + \frac{D}{c_1} \ln \frac{v_1 - \dfrac{c_1}{\rho_{1i}} - \dfrac{c_2}{\rho_{2i}}}{v_{10} - \dfrac{c_1}{\rho_{1i}} - \dfrac{c_2}{\rho_{2i}}} = - \frac{K}{c_1} \int\limits_0^x f(x)\,dx,$$

which may be written as

$$\left[\frac{u_1 f(x)}{u_{10} f(x_0)} \right]^{\frac{A}{c_1}} e^{B_1 \frac{u_1 f(x) - u_{10} f(x_0)}{u_1 f(x) u_{10} f(x_0)}} \left[\frac{u_1 f(x) - \dfrac{c_1}{\rho_{1i}}}{u_{10} f(x_0) - \dfrac{c_1}{\rho_{1i}}} \right]^{-\frac{c_2}{c_1}} \left[\frac{u_1 f(x) - \dfrac{c_1}{\rho_{1i}} - \dfrac{c_2}{\rho_{2i}}}{u_{10} f(x_0) - \dfrac{c_1}{\rho_{1i}} - \dfrac{c_2}{\rho_{2i}}} \right]^{\frac{D}{c_1}} = e^{-\frac{K}{c_1} \int\limits_{x_0}^x f(x)\,dx} \qquad 1.5)$$

Proceeding analogously (under the condition that $v_1 = v_{20} = u_{20} f(x)$ is satisfied for $x = x_0$), we have the following expression to determine the velocity of the second phase:

$$\left[\frac{u_2 f(x)}{u_{20} f(x)} \right]^{\frac{A}{c_1}} e^{B_2 \frac{u_1 f(x) - u_{20} f(x)}{u_2 f(x) u_{20} f(x)}} \left[\frac{u_2 f(x) - \dfrac{c_2}{\rho_{2i}}}{u_{20} f(x_0) - \dfrac{c_2}{\rho_{2i}}} \right]^{-\frac{c_1}{c_1}} \left[\frac{u_2 f(x) - \dfrac{c_1}{\rho_{1i}} - \dfrac{c_2}{\rho_{2i}}}{u_{20} f(x_0) - \dfrac{c_1}{\rho_{1i}} - \dfrac{c_2}{\rho_{2i}}} \right] = e^{-\frac{K}{c_1} \int\limits_{x_0}^x f(x)\,dx}, \qquad (11.6)$$

where

$$A = \frac{c_1 \dfrac{c_1}{\rho_{1i}} \left(\dfrac{c_2}{\rho_{1i}} + 2 \dfrac{c_2}{\rho_{2i}} \right) + c_2 \dfrac{c_2}{\rho_{2i}} \left(2 \dfrac{c_1}{\rho_{1i}} + \dfrac{c_2}{\rho_{2i}} \right)}{\left(\dfrac{c_1}{\rho_{1i}} + \dfrac{c_2}{\rho_{2i}} \right)^2}$$

$$B_1 = \frac{\dfrac{c_2^2}{\rho_{1i}\rho_{2i}} - \left(\dfrac{c_1}{\rho_{1i}} \right)^2}{\left(\dfrac{c_1}{\rho_{1i}} + \dfrac{c_2}{\rho_{2i}} \right)}, \qquad D = \frac{\left(\dfrac{c_1}{\rho_{1i}} \right)^2 c_2 + \left(\dfrac{c_2}{\rho_{2i}} \right)^2 c_1}{\left(\dfrac{c_1}{\rho_{1i}} + \dfrac{c_2}{\rho_{2i}} \right)^2} \qquad (11.7)$$

$$B_2 = \frac{\dfrac{c_1^2}{\rho_{1i}\rho_{2i}} - \left(\dfrac{c_2}{\rho_{2i}} \right)^2}{\dfrac{c_1}{\rho_{1i}} + \dfrac{c_2}{\rho_{2i}}}$$

We have obtained expressions for u_1, u_2 as implicit functions of x. To find f_1 and f_2 it is necessary to substitute the values of u_1, u_2 from (11.2) into (11.3); after simple manipulations we have

$$\frac{\dfrac{c_1^2}{\rho_{1i}} (f_1 - 1)^3 - \dfrac{c_2^2}{\rho_{2i}} f_1^3}{f_1 (f_1 - 1) \left[\left(\dfrac{c_1}{\rho_{1i}} + \dfrac{c_2}{\rho_{2i}} \right) f_1 - \dfrac{c_1}{\rho_{1i}} \right]} \frac{df_1}{dx} = - K f(x). \qquad (11.8)$$

Decomposing the rational fraction in the left side of this last equation into elementary fractions, we write

$$\left[\frac{1}{f_1} + \frac{\frac{c_1}{c_2}}{f_1 - 1} + \frac{\left(\frac{c_1}{\rho_{1i}}\right)^2 c_2 + \left(\frac{c_2}{\rho_{2i}}\right)^2 c_1}{\left(\frac{c_1}{\rho_{1i}} + \frac{c_2}{\rho_{2i}}\right)^2 c_1} \quad \frac{1}{\left(\frac{c_1}{\rho_{1i}} + \frac{c_2}{\rho_{2i}}\right) f_1 - \frac{c_1}{\rho_{1i}}} + \frac{\frac{c_2}{\rho_{2i}} c_2 - \frac{c_1}{\rho_{1i}} c_1}{\frac{c_1}{\rho_{1i}} + \frac{c_2}{\rho_{2i}}}\right] \frac{df_1}{dx} = \frac{K}{c_1} f(x).$$

Integration of this equation under the condition $f_1 = f_{10}$ when $x = x_0$ yields an expression for the porosity:

$$\ln\left[\frac{f_1}{f_{10}}\left(\frac{f_1 - 1}{f_{10} - 1}\right)^{\frac{c_2}{c_1}}\right] + \frac{\left(\frac{c_1}{\rho_{1i}}\right)^2 c_2 + \left(\frac{c_2}{\rho_{2i}}\right) c_1}{\left(\frac{c_1}{\rho_{1i}} + \frac{c_2}{\rho_{2i}}\right)^2 c_1} \ln\left[\frac{\left(\frac{c_1}{\rho_{1i}} + \frac{c_2}{\rho_{2i}}\right) f_1 - \frac{c_1}{\rho_{1i}}}{\left(\frac{c_1}{\rho_{1i}} + \frac{c_2}{\rho_{2i}}\right) f_{10} - \frac{c_1}{\rho_{1i}}}\right] + \frac{\frac{c_2}{\rho_{2i}} c_2 - \frac{c_1}{\rho_{1i}} c_1}{\left(\frac{c_1}{\rho_{1i}} + \frac{c_2}{\rho_{2i}}\right) c_1}(f_1 - f_{10}) = \frac{K}{c_1}\int_{x_0}^{x} f(x)\,dx.$$

$$(11.9)$$

Analogously, under the condition $f_2 = f_{20}$ when $x = x_0$, we have for f_2

$$\ln\left[\frac{f_2}{f_{20}}\left(\frac{f_2 - 1}{f_{20} - 1}\right)^{\frac{c_1}{c_2}}\right] + \frac{\left(\frac{c_1}{\rho_{1i}}\right)^2 c_2 + \left(\frac{c_2}{\rho_{2i}}\right)^2 c_1}{\left(\frac{c_1}{\rho_{1i}} + \frac{c_2}{\rho_{2i}}\right)^2 c_2} \ln\left[\frac{\left(\frac{c_1}{\rho_{1i}} + \frac{c_2}{\rho_{2i}}\right) f_2 - \frac{c_2}{\rho_{2i}}}{\left(\frac{c_1}{\rho_{1i}} + \frac{c_2}{\rho_{2i}}\right) f_{20} - \frac{c_2}{\rho_{2i}}}\right] + \frac{\frac{c_1^2}{\rho_{1i}} - \frac{c_2^2}{\rho_{2i}}}{\left(\frac{c_1}{\rho_{1i}} + \frac{c_2}{\rho_{2i}}\right) c_2}(f_2 - f_{20}) = \frac{K}{c_2}\int_{x_0}^{x} f(x)\,dx.$$

$$(11.10)$$

The pressure is determined as follows. Addition of the equations in the system (10.1) yields

$$\frac{dp}{dx} = -\frac{c_1}{f(x)}\frac{du_1}{dx} - \frac{c_2}{f(x)}\frac{du_2}{dx},$$

from which

$$\frac{dp}{dx} = -\frac{d}{dx}\left[\frac{c_1 u_1 + c_2 u_2}{f(x)}\right] + \left(\frac{c_1^2}{\rho_1} + \frac{c_2^2}{\rho_2}\right)\frac{f'(x)}{f^2(x)}.$$

Considering the last member small and discarding it, then integrating under the condition $p = p_0$ when $x = x_0$, we will have

$$p - p_0 = \frac{c_1 u_{10} + c_2 u_{20}}{f(x_0)} - \frac{c_1 u_1 + c_2 u_2}{f(x)}.$$

$$(11.11)$$

Let us note that (throughout this section)

$$f(x) = f(x_0) + (x - x_0)\tan\alpha$$

if the diffusor is plane;

$$f(x) = f(x_0) + \pi(x^2 - x_0^2)\tan^2\alpha$$

if the diffusor is conical. There the formulas obtained are true if terms containing the small quantities $f'(x)/f^2(x)$ can be discarded from the equations. As can be seen,

$$\frac{f'(x)}{f^2(x)} = \frac{1}{x^2 \tan\alpha}$$

for a plane conduit;

$$\frac{f'(x)}{f^2(x)} = \frac{2}{\pi x^3 \tan^2\alpha}$$

for a conical conduit. Therefore, for given α the preceding equations become small starting with some value of x.

Hence, the solutions obtained in this section are valid for the values $x_0 \le x \le \infty$ or $x_0 < x \le \infty$ depending on where the initial section has been taken.

Let us note that for small values of x_0 the solutions of the problem of two-phase ideal incompressible media motion in expanding conduits (for a small angle α) in the linear and the nonlinear formulations considered here complement each other in the sense that the former is valid near the initial section, while the latter is valid far from the initial section.

On the basis of the formulas obtained, let us consider changes in the parameters along the flow. For

$$x = x_0 e^{-\frac{K}{c_1} \int_{x_0}^{x} f(x)\, dx}$$

the right side of (11.5) becomes unity. By verifying the left side it is easy to see that each member of the product also becomes unity. This is valid for (11.6) also.

Since $f(x)$ is a continuously increasing function, then $\int_{x_0}^{x} f(x)\, dx \to \infty$ as $x \to \infty$. Hence, the left sides of (11.5) and (11.6) tend to zero. This is possible under the following conditions:

1) $\quad \dfrac{u_1 f(x)}{u_{10} f(x_0)} \to 0,$

2) $\quad e^{B_1 \frac{u_1 f(x) - u_{10} f(x_0)}{u_1 f(x)\, u_{10} f(x_0)}} \to 0,$

3) $\quad \left[\dfrac{u_1 f(x) - \dfrac{c_1}{\rho_{1i}}}{u_{10} f(x_0) - \dfrac{c_1}{\rho_{1i}}}\right]^{-\frac{c_2}{c_1}} \to 0,$

4) $\quad \dfrac{u_1 f(x) - \dfrac{c_1}{\rho_{1i}} - \dfrac{c_2}{\rho_{2i}}}{u_{10} f(x_0) - \dfrac{c_1}{\rho_{1i}} - \dfrac{c_2}{\rho_{2i}}} \to 0.$

The first condition is possible when $u_1 f(x) = 0$; however it here follows from (11.2) that $f_1 \to \infty$, $f_2 \to \infty$, which is impossible. The second condition yields $u_1 f(x) = 0$ or $u_1 f(x) = \infty$, depending on the sign of B_1; we then have $f_1 \to \infty$ or $f_1 = 0$, which does not satisfy the condition of a two-phase flow. The above is also true for the third condition. The fourth condition yields

$$\lim_{x \to \infty} u_1 f(x) = \frac{c_1}{\rho_{1i}} + \frac{c_2}{\rho_{2i}}. \tag{11.12}$$

Completely analogously we determine

$$\lim_{x \to \infty} u_2 f(x) = \frac{c_1}{\rho_{1i}} + \frac{c_2}{\rho_{2i}}. \tag{11.13}$$

Since $f(x)$ is an increasing function, we have $u_1 \to 0$, $u_2 \to 0$ as $x \to \infty$ from (11.12) and (11.13). In this case the inequalities

$$u_1 f(x) > (f_{10} u_{10} + f_{20} u_{20}) f(x_0) > u_2 f(x) \qquad (11.14)$$

for $u_{10} > u_{20}$

and

$$u_1 f(x) < (f_{10} u_{10} + f_{20} u_{20}) f(x_0) < u_2 f(x) \qquad (11.15)$$

for $u_{10} < u_{20}$ are satisfied for finite x.

To prove the validity of the left side of (11.14), $v_1 > L$ for $u_{10} > u_{20}$, where

$$L = (f_{10} u_{10} + f_{20} u_{20}) f(x_0),$$

we proceed as follows. Let v_1 have a value less than $L (v_1 < L)$. In this case, since $v_{10} > L$†
in the initial section and v_1 is a decreasing function, a point a exists at which the condition
$v_1 = L$ is satisfied, and near this point, i.e., at $x = a + \varepsilon$, the inequality

$$v_1 < L \frac{dv_1}{dx} < 0 \qquad (11.16)$$

is valid.

After some manipulation we can obtain from (11.4)

$$\frac{s}{v_1 - L} \frac{dv_1}{dx} = -f(x); \qquad (11.17)$$

where

$$s = \frac{f_2^3 v_1^2 + f_1 \dfrac{c_2^2}{\rho_{1i}\,\rho_{2i}}}{f_2\, v_1^2} c_1.$$

Evidently s and $f(x)$ are always positive; hence the right side of (11.17) will be positive
at the point $x = a + \varepsilon$ and the left side will be negative. The resulting contradiction asserts
that

$$u_1 f(x) > (f_{10} u_{10} + f_{20} u_{20}) f(x_0).$$

It may analogously be proved that

$$u_2 f(x) < (f_{10} u_{10} + f_{20} u_{20}) f(x_0).$$

The validity of inequality (11.14) is thereby proved. The validity of inequality (11.15) is proved
in the same manner.

The following deductions may be made on the basis of the above:

1. The velocity of the phases whose initial velocity is greater than the initial velocity of
the other phase is always greater than the function

$$(f_{10} u_{10} + f_{20} u_{20}) \frac{f(x_0)}{f(x)}, \qquad (11.18)$$

---

† Multiplying the left and right sides of the equality $f_{10} u_{10} + f_{20} u_{20} = u_{10} - f_{20}(u_{10} - u_{20}) < u_{10}$ by
$f(x_0)$, we obtain $u_{10} f(x_0) > (f_{10} u_{10} + f_{20} u_{20}) f(x_0)$ for $u_{10} > u_{20}$.

and the velocity of the phase whose initial velocity is less is always less than the function (11.18), and tends to zero as $x \to \infty$.

2. The porosity of the phase with the greater initial velocity decreases, but that of the phase with the lesser velocity increases, each tending to a constant:

$$\lim_{x \to \infty} f_1 = \frac{f_{10} u_{10}}{f_{10} u_{10} + f_{20} u_{20}},$$

$$\lim_{x \to \infty} f_2 = \frac{f_{20} u_{20}}{f_{10} u_{10} + f_{20} u_{20}}.$$

3. The pressure tends to the constant value (11.11):

$$p = p_0 + \frac{c_1 u_{10} + c_2 u_{20}}{f(x_0)}$$

If we put

$$f(x) = f(x_0) = \text{const}$$

in the formulas in this section, we will then have the case of motion in constant-section conduits.

§12. Steady Motion of Three-Phase Compressible Ideal Media in Diverging Inclined Conduits

Let there be steady-state motion of compressible ideal three-phase media in a constant-section conduit whose axis makes an angle ε with the vertical. We direct the 0x axis along the conduit axis. Then (7.3), (7.4), (7.5), (3.2), (2.1), and (7.2) after linearization, become

$$\left.\begin{aligned}
\rho_{10} u_{10} \frac{du'_1}{dx} + (K_{12} + K_{13}) u'_1 - K_{12} u'_2 - K_{13} u'_3 = \\
= -\frac{\rho_{10}}{\rho_{1i}^0} \frac{dp'}{dx} - (K_{12} + K_{13}) u_{10} + K_{12} u_{20} + K_{13} u_{30} - g \cos \varepsilon \\[1em]
\rho_{20} u_{20} \frac{du'_2}{dx} - K_{12} u'_1 + (K_{12} + K_{23}) u'_2 - K_{23} u'_3 = \\
= -\frac{\rho_{20}}{\rho_{2i}^0} \frac{dp'}{dx} + K_{12} u_{10} - (K_{12} + K_{23}) u_{20} + K_{23} u_{30} - g \cos \varepsilon \\[1em]
\rho_{30} u_{30} \frac{du'_3}{dx} - K_{13} u'_1 - K_{23} u'_2 + (K_{13} + K_{23}) u'_3 = \\
= -\frac{\rho_{30}}{\rho_{3i}^0} \frac{dp'}{dx} + K_{13} u_{10} + K_{23} u_{20} - (K_{13} - K_{23}) u_{30} - g \cos \varepsilon
\end{aligned}\right\} \quad (12.1)$$

$$\left.\begin{aligned}
\rho_{10} u_{10} f' + \rho_{10} f_0 u'_1 + u_{10} f_0 \rho'_1 = 0 \\
\rho_{20} u_{20} f' + \rho_{20} f_0 u'_2 + u_{20} f_0 \rho'_2 = 0 \\
\rho_{30} u_{30} f' + \rho_{30} f_0 u'_3 + u_{30} f_0 \rho'_3 = 0
\end{aligned}\right\} \quad (12.2)$$

$$\frac{\rho'_1}{\rho_{1i}^0} + \frac{\rho'_2}{\rho_{2i}^0} + \frac{\rho'_3}{\rho_{3i}^0} - \xi p' = 0, \quad (12.3)$$

$$\left.\begin{aligned}
p' = a_{10}^2 \rho'_{1i}, \qquad p' = a_{20}^2 \rho'_{2i} \\
p' = a_{30}^2 \rho'_{3i}
\end{aligned}\right\} \quad (12.4)$$

where

$$\xi = \frac{\rho_{10}}{\rho_{1i}^{02} a_{10}^2} + \frac{\rho_{20}}{\rho_{2i}^{02} a_{20}^2} + \frac{\rho_{30}}{\rho_{3i}^{02} a_{30}^2}.$$

To solve the system of equations (12.1) - (12.4), let us consider it first without the right sides. Let us make the substitutions

$$u_1' = \gamma_1 e^{\lambda x}, \quad u_2' = \gamma_2 e^{\lambda x}, \quad u_3' = \gamma_3 e^{\lambda x}.$$

We then have

$$(\rho_{10} u_{10} \lambda + K_{12} + K_{13}) \gamma_1 - K_{12} \gamma_2 - K_{13} \gamma_3 = 0,$$
$$-K_{12} \gamma_1 + (\rho_{20} u_{20} \lambda + K_{12} + K_{23}) \gamma_2 - K_{23} \gamma_3 = 0,$$
$$-K_{13} \gamma_1 - K_{23} \gamma_2 + (\rho_{30} u_{30} \lambda + K_{13} + K_{23}) \gamma_3 = 0.$$

It is known that the necessary and sufficient condition that this system have a nontrivial solution is

$$\Delta = \begin{vmatrix} (\rho_{10} u_{10}\lambda + K_{12} + K_{13}) & -K_{12} & -K_{13} \\ -K_{12} & (\rho_{20} u_{20}\lambda + K_{12} + K_{13}) & -K_{23} \\ -K_{13} & -K_{23} & (\rho_{30}u_{30}\lambda + K_{13} + K_{23}) \end{vmatrix} = 0,$$

which is the characteristic equation of the system. Expanding it, we obtain a third-degree algebraic equation in λ whose roots will be

$$\lambda_1 = 0, \quad \lambda_2 = -\alpha_1, \quad \lambda_3 = -\alpha_2,$$

where

$$\alpha_1 = \frac{1}{2}\left(\frac{K_{12}+K_{13}}{\rho_{10} u_{10}} + \frac{K_{12}+K_{23}}{\rho_{20} u_{20}} + \frac{K_{13}+K_{23}}{\rho_{30} u_{30}} \right) - \sqrt{ \frac{1}{4}\left(\frac{K_{12}+K_{13}}{\rho_{10} u_{10}} + \frac{K_{12}+K_{23}}{\rho_{20} u_{20}} + \frac{K_{13}+K_{23}}{\rho_{30} u_{30}} \right)^2 - K \frac{\rho_{10}u_{10} + \rho_{20}u_{20} + \rho_{30}u_{30}}{\rho_{10}\rho_{20}\rho_{30}u_{10}u_{20}u_{30}} },$$

$$\alpha_2 = \frac{1}{2}\left(\frac{K_{12}+K_{13}}{\rho_{10} u_{10}} + \frac{K_{12}+K_{23}}{\rho_{20} u_{20}} + \frac{K_{13}+K_{23}}{\rho_{30} u_{30}} \right) + \sqrt{ \frac{1}{4}\left(\frac{K_{12}+K_{13}}{\rho_{10} u_{10}} + \frac{K_{12}-K_{23}}{\rho_{20} u_{20}} + \frac{K_{13}+K_{23}}{\rho_{30} u_{30}} \right)^2 - K \frac{\rho_{10}u_{10} + \rho_{20}u_{20} + \rho_{30}u_{30}}{\rho_{10}\rho_{20}\rho_{30}u_{10}u_{20}u_{30}} },$$

$$K = K_{12} K_{13} + K_{12} K_{23} + K_{13} K_{23}.$$

A system of two independent equations in γ_1, γ_2, γ_3 corresponds to the root $\lambda_1 = 0$; for example,

$$(K_{12} + K_{13}) \gamma_1^{(1)} - K_{12} \gamma_2^{(1)} - K_{13} \gamma_3^{(1)} = 0,$$
$$-K_{12} \gamma_1^{(1)} + (K_{12} + K_{23}) \gamma_2^{(1)} - K_{23} \gamma_3^{(1)} = 0,$$

from which we obtain

$$\frac{\gamma_1^{(1)}}{\begin{vmatrix} -K_{12} & -K_{13} \\ (K_{12}+K_{23}) & -K_{23} \end{vmatrix}} = \frac{\gamma_2^{(1)}}{\begin{vmatrix} -K_{13} & (K_{12}+K_{13}) \\ -K_{23} & -K_{12} \end{vmatrix}} = \frac{\gamma_3^{(1)}}{\begin{vmatrix} (K_{12}+K_{13}) & -K_{12} \\ -K_{12} & (K_{12}+K_{23}) \end{vmatrix}},$$

or

$$\gamma_1^{(1)} = \gamma_2^{(1)} = \gamma_3^{(1)} = -c (K_{12} K_{13} + K_{12} K_{23} + K_{13} K_{23}) = cK = c'.$$

Then one of the particular solutions of the homogeneous system will be

$$u_1'^{(1)} = u_2'^{(1)} = u_3'^{(1)} = c' = \text{const.}$$

The system

$$(-\rho_{10}\,u_{10}\,\alpha_1 + K_{12} + K_{13})\,\gamma_1^{(2)} - K_{12}\,\gamma_2^{(2)} - K_{13}\,\gamma_3^{(2)} = 0,$$

$$-K_{12}\,\gamma_1^{(2)} + (-\rho_{20}\,u_{20}\,\alpha_1 + K_{12} + K_{23})\,\gamma_2^{(2)} - K_{23}\,\gamma_3^{(2)} = 0,$$

for example, corresponds to the root $\lambda = \lambda_2 = -\alpha_1$, and it yields

$$\frac{\gamma_1^{(2)}}{\begin{vmatrix} -K_{12} & -K_{13} \\ (-\rho_{20}u_{20}\alpha_1+K_{12}+K_{13}) & -K_{23} \end{vmatrix}} = \frac{\gamma_2^{(2)}}{\begin{vmatrix} -K_{13} & (-\rho_{10}u_{10}\alpha_1+K_{12}+K_{13}) \\ -K_{23} & -K_{12} \end{vmatrix}} = \frac{\gamma_3^{(2)}}{\begin{vmatrix} (\rho_{10}u_{10}\alpha_1+K_{12}+K_{13}) & -K_{12} \\ -K_{12} & (-\rho_{20}u_{20}\alpha_1+K_{12}+K_{23}) \end{vmatrix}} = c'',$$

on the basis of which we obtain the particular solutions

$$u_1^{'(2)} = \gamma_1^{(2)}\,e^{\lambda_2 x} = c''\,(K - K_{13}\,\rho_{20}\,u_{20}\,\alpha_1)\,e^{-\alpha_1 x}$$

$$u_2^{'(2)} = \gamma_2^{(2)}\,e^{\lambda_2 x} = c''\,(K - K_{23}\,\rho_{10}\,u_{10}\,\alpha_1)\,e^{-\alpha_1 x},$$

$$u_3^{'(2)} = \gamma_3^{(2)}\,e^{\lambda_2 x} = c''\big\{K - [(K_{12}+K_{23})\,\rho_{10}\,u_{10} + (K_{12}+K_{13})\,\rho_{20}\,u_{20}]\,\alpha_1 + \rho_{10}\,\rho_{20}\,u_{10}\,u_{20}\,\alpha_1^2\big\}e^{-\alpha_1 x}.$$

In exactly the same way, the particular solutions

$$u_1^{'(3)} = \gamma_1^{(3)}\,e^{\lambda_3 x} = c'''\,(K - K_{13}\,\rho_{20}\,u_{20}\,\alpha_2)\,e^{-\alpha_2 x},$$

$$u_2^{'(3)} = \gamma_2^{(3)}\,e^{\lambda_3 x} = c'''\,(K - K_{23}\,\rho_{10}\,u_{10}\,\alpha_2)\,e^{-\alpha_2 x},$$

$$u_3^{'(3)} = \gamma_3^{(3)}\,e^{\lambda_3 x} = c'''\big\{K - [(K_{12}-K_{23})\,\rho_{10}\,u_{10} + (K_{12}+K_{13})\,\rho_{20}\,u_{20}]\,\alpha_2 + \rho_{10}\,\rho_{20}\,u_{10}\,u_{20}\,\alpha_2^2\big\}e^{-\alpha_2 x}$$

correspond to the root $\lambda = \lambda_3 = -\alpha_2$, where the c', c", c''' are arbitrary constants.

Therefore, the general solution of the homogeneous system will be

$$\left.\begin{aligned} u_1^{'} &= c_1 + c_2\,(K - K_{13}\,\rho_{20}\,u_{20}\,\alpha_1)\,e^{-\alpha_1 x} + c_3\,(K + K_{13}\,\rho_{20}\,u_{20}\,\alpha_2)\,e^{-\alpha_2 x} \\ u_2^{'} &= c_1 + c_2\,(K - K_{23}\,\rho_{10}\,u_{10}\,\alpha_1)\,e^{-\alpha_1 x} + c_3\,(K - K_{23}\,\rho_{10}\,u_{10}\,\alpha_2)\,e^{-\alpha_2 x} \\ u_3^{'} &= c_1 + c_2\,(K - M\alpha_1 + N\alpha_1^2)\,e^{-\alpha_1 x} + c_3\,(K - M\alpha_2 + N\alpha_2^2)\,e^{-\alpha_2 x} \end{aligned}\right\} \tag{12.5}$$

where

$$M = (K_{12} + K_{23})\,\rho_{10}\,u_{10} + (K_{12} + K_{13})\,\rho_{20}\,u_{20};$$
$$N = \rho_{10}\,\rho_{20}\,u_{10}\,u_{20}.$$

We find the solution of the inhomogeneous system (12.1) – (12.4) by the method of variation of constants. To do this, by substituting (12.5) into this system we consider c_1, c_2, c_3 to be variables depending on x. Separating the resulting equations with respect to $\rho_{10}\,u_{10}$, $\rho_{20}\,u_{20}$, and $\rho_{30}\,u_{30}$, respectively, we have

$$\left.\begin{aligned} \frac{dc_1}{dx} + (K + K_{23}\,\rho_{10}\,u_{10}\,\alpha_1)\,e^{-\alpha_1 x}\frac{dc_2}{dx} + (K - K_{23}\,\rho_{10}\,u_{10}\,\alpha_2)\,e^{-\alpha_2 x}\frac{dc_3}{dx} &= -\frac{1}{\rho_{2i}^0\,u_{20}}\frac{dp'}{dx} + L_2 \\ \frac{dc_1}{dx} + (K - M\alpha_1 + N\alpha_1^2)\,e^{-\alpha_1 x}\frac{dc_2}{dx} + (K - M\alpha_2 + N\alpha_2^2)\,e^{-\alpha_2 x}\frac{dc_3}{dx} &= -\frac{1}{\rho_{3i}^0\,u_{10}}\frac{dp'}{dx} + L_3 \end{aligned}\right\} \tag{12.6}$$

where

$$L_1 = \frac{1}{\rho_{10} u_{10}}\Big[-(K_{12} + K_{13})\, u_{10} + K_{12}\, u_{20} + K_{13}\, u_{30} - g \cos\varepsilon\Big],$$

$$L_2 = \frac{1}{\rho_{20} u_{20}}\Big[K_{12}\, u_{10} - (K_{12} + K_{23})\, u_{20} - K_{23}\, u_{30} - g \cos\varepsilon\Big],$$

$$L_3 = \frac{1}{\rho_{30} u_{30}}\Big[K_{13}\, u_{10} + K_{23}\, u_{20} - (K_{13} + K_{23})\, u_{30} - g \cos\varepsilon\Big].$$

The resulting system of inhomogeneous linear equations is solvable for dc_1/dx, dc_2/dx, dc_3/dx since its determinant D is not zero, and the system $u_j^{'(n)}$ (j, n = 1, 2, 3) is fundamental [7]. Hence, we may find the solution of the system (12.6) by means of the Cramer formula. To this end, let us write the determinant D as

$$D = \begin{vmatrix} 1 & (K - K_{13}\rho_{20} u_{20}\,\alpha_1)\, e^{-\alpha_1 x} & (K - K_{13}\rho_{20} u_{20}\,\alpha_2)\, e^{-\alpha_2 x} \\ 1 & (K - K_{23}\rho_{10} u_{10}\,\alpha_1)\, e^{-\alpha_1 x} & (K - K_{23}\rho_{20} u_{20}\,\alpha_2)\, e^{-\alpha_2 x} \\ 1 & (K - M\alpha_1 + N\alpha_1^2)\, e^{-\alpha_1 x} & (K - M\alpha_2 + N\alpha_2^2)\, e^{-\alpha_2 x} \end{vmatrix} = \begin{vmatrix} 1 & f_{12} & f_{13} \\ 1 & f_{22} & f_{23} \\ 1 & f_{32} & f_{33} \end{vmatrix} \cdot e^{-\alpha_1 x} \cdot e^{-\alpha_2 x} = D_0\, e^{-(\alpha_1 + \alpha_2)x} \neq 0,$$

where f_{jk}, j = 1, 2, 3 (k = 2, 3) denotes the factors in parentheses and D_0 denotes the last determinant. Then the determinants D_1, D_2, D_3 will be

$$D_1 = \left(-D_{11} \frac{dp'}{dx} + D_{12}\right) e^{-(\alpha_1 + \alpha_2)x},$$

$$D_2 = \left(-D_{21} \frac{dp'}{dx} + D_{22}\right) e^{-\alpha_1 x},$$

$$D_3 = \left(-D_{31} \frac{dp'}{dx} + D_{32}\right) e^{-\alpha_1 x};$$

where

$$D_{11} = \begin{vmatrix} \dfrac{1}{\rho_{1i}^{0} u_{10}} & f_{12} & f_{13} \\[2ex] \dfrac{1}{\rho_{2i}^{0} u_{20}} & f_{22} & f_{23} \\[2ex] \dfrac{1}{\rho_{3i}^{0} u_{30}} & f_{32} & f_{33} \end{vmatrix},$$

$$D_{12} = \begin{vmatrix} L_1 & f_{12} & f_{13} \\ L_2 & f_{22} & f_{23} \\ L_3 & f_{23} & f_{33} \end{vmatrix},$$

$$D_{21} = \begin{vmatrix} 1 & \dfrac{1}{\rho_{1i}^{0} u_{10}} & f_{13} \\[2ex] 1 & \dfrac{1}{\rho_{2i}^{0} u_{20}} & f_{23} \\[2ex] 1 & \dfrac{1}{\rho_{3i}^{0} u_{30}} & f_{33} \end{vmatrix},$$

$$D_{22} = \begin{vmatrix} 1 & L_1 & f_{13} \\ 1 & L_2 & f_{23} \\ 1 & L_3 & f_{33} \end{vmatrix},$$

$$D_{31} = \begin{vmatrix} 1 & f_{12} & \dfrac{1}{p^0_{1i}\,u_{10}} \\ 1 & f_{23} & \dfrac{1}{p^0_{2i}\,u_{20}} \\ 1 & f_{33} & \dfrac{1}{p^0_{3i}\,u_{30}} \end{vmatrix},$$

$$D_{32} = \begin{vmatrix} 1 & f_{12} & L_1 \\ 1 & f_{22} & L_2 \\ 1 & f_{32} & L_3 \end{vmatrix}.$$

Therefore,

$$\frac{dc_1}{dx} = \frac{D_1}{D} = -\frac{D_{11}}{D_0}\frac{dp'}{dx} + \frac{D_{12}}{D_0},$$

$$\frac{dc_2}{dx} = \frac{D_2}{D} = \left(-\frac{D_{21}}{D_0}\frac{dp'}{dx} + \frac{D_{22}}{D_0}\right) e^{\alpha_1 x},$$

$$\frac{dc_3}{dx} = \frac{D_3}{D} = \left(-\frac{D_{31}}{D_0}\frac{dp'}{dx} + \frac{D_{32}}{D}\right) e^{\alpha_2 x}.$$

Since p' = 0 for x = 0, by integrating the last equations we obtain

$$c_1 = -\frac{D_{11}}{D_0}p' + \frac{D_{12}}{D_0}x,$$

$$c_2 = -\frac{D_{21}}{D_0}\int_0^x e^{\alpha_1 z}\frac{dp'}{dz}\,dz + \frac{D_{22}}{D_0\,\alpha_1}(e^{\alpha_1 x} - 1),$$

$$c_3 = -\frac{D_{31}}{D_0}\int_0^x e^{\alpha_2 z}\frac{dp'}{dz}\,dz + \frac{D_{32}}{D_0\,\alpha_2}(e^{\alpha_2 x} - 1).$$

Substituting these values of c_1, c_2, c_3 in (12.5) yields the solution of the inhomogeneous system in the form

$$'u = -a_{11}p' + a_{12}x + a_{13}(1 - e^{-\alpha_1 x}) + a_{14}(1 - e^{-\alpha_2 x}) - a_{15}\int_0^x e^{-\alpha_1(x-z)}\frac{dp'}{dz}\,dz - a_{16}\int_0^x e^{-\alpha_2(x-z)}\frac{dp'}{dz}\,dz$$

$$u_2' = -a_{11}p' + a_{12}x + a_{23}(1 - e^{-\alpha_1 x}) + a_{24}(1 - e^{-\alpha_2 x}) - a_{25}\int_0^x e^{-\alpha_1(x-z)}\frac{dp'}{dz}\,dz - a_{26}\int_0^x e^{-\alpha_2(x-z)}\frac{dp'}{dz}\,dz \qquad \Big\} ,\ (12.7)$$

$$u_3' = -a_{11}p' + a_{12}x + a_{33}(1 - e^{-\alpha_1 x}) + a_{34}(1 - e^{-\alpha_2 x}) - a_{35}\int_0^x e^{-\alpha_1(x-z)}\frac{dp'}{dz}\,dz - a_{36}\int_0^x e^{-\alpha_2(x-z)}\frac{dp'}{dz}\,dz$$

where

$$a_{11} = \frac{D_{11}}{D_0}, \quad a_{12} = \frac{D_{12}}{D_0}, \quad a_{13} = \frac{D_{22}}{D_0}\frac{f_{12}}{\alpha_1}, \quad a_{24} = \frac{D_{33}}{D_0}\frac{f_{13}}{\alpha_2},$$

$$a_{15} = \frac{D_{21}}{D_0} f_{12}, \ a_{16} = \frac{D_{31}}{D_0} f_{13}, \ a_{23} = \frac{D_{22}}{D_0} \frac{f_{22}}{a_1}, \ a_{24} = \frac{D_{32}}{D_0} \frac{f_{23}}{a_2},$$

$$a_{25} = \frac{D_{21}}{D_0} f_{22}, \ a_{26} = \frac{D_{31}}{D_0} f_{23}, \ a_{33} = \frac{D_{22}}{D_0} \frac{f_{32}}{a_1}, \ a_{34} = \frac{D_{32}}{D_0} \frac{f_{33}}{a_2},$$

$$a_{35} = \frac{D_{21}}{D_0} f_{32}, \ a_{36} = \frac{D_{31}}{D_0} f_{33}.$$

Substituting the values of ρ_1', ρ_2', ρ_3' from (12.2) into (12.3), and taking into account that $f_1/f_0 = \eta x$, we have

$$-\xi p' = \frac{\rho_{10}}{\rho_{1i}^0 u_{10}} u_1' + \frac{\rho_{20}}{\rho_{2i}^0 u_{20}} u_2' + \frac{\rho_{30}}{\rho_{3i}^0 u_{30}} u_3' - \eta x. \tag{12.8}$$

Replacing u_1', u_2', u_3' by their values from the system (12.7), we obtain

$$-\xi p' = -\left(\frac{\rho_{10}}{\rho_{1i}^0 u_{10}} + \frac{\rho_{20}}{\rho_{2i}^0 u_{20}} + \frac{\rho_{30}}{\rho_{3i}^0 u_{30}}\right) a_{11} p' + \left[\left(\frac{\rho_{10}}{\rho_{1i}^0 u_{10}} + \frac{\rho_{20}}{\rho_{3i}^0 u_{20}} + \frac{\rho_{30}}{\rho_{3i}^0 u_{30}}\right) a_{12} - \eta\right] x +$$

$$+ \left(\frac{\rho_{10}^0}{\rho_{1i}^0 u_{10}} a_{13} + \frac{\rho_{20}}{\rho_{2i}^0 u_{20}} a_{23} + \frac{\rho_{30}}{\rho_{3i}^0 u_{30}} a_{33}\right)\left(1 - e^{-a_1 x}\right) + \left(\frac{\rho_{10}}{\rho_{1i}^0 u_{10}} a_{11} + \frac{\rho_{20}}{\rho_{2i}^0 u_{20}} a_{21} + \frac{\rho_{30}}{\rho_{3i}^0 u_{30}} a_{31}\right)\left(1 - e^{-a_2 x}\right) -$$

$$- \left(\frac{\rho_{10}}{\rho_{1i}^0 u_{10}} a_{15} + \frac{\rho_{20}}{\rho_{2i}^0 u_{20}} a_{25} + \frac{\rho_{30}}{\rho_{3i}^0 u_{30}} a_{35}\right)\int_0^x e^{-a_1(x-z)} \frac{dp'}{dz} dz -$$

$$- \left(\frac{\rho_{10}}{\rho_{1i}^0 u_{10}} a_{16} + \frac{\rho_{20}}{\rho_{2i}^0 u_{20}} a_{26} + \frac{\rho_{30}}{\rho_{3i}^0 u_{30}} a_{36}\right)\int_0^x e^{-a_1(x-z)} \frac{dp'}{dz} dz.$$

Let us transfer the first member on the right side over to the left, and let c denote the coefficient of p', and A_1, A_2, A_3, A, and B the coefficients on the right side. Then

$$cp' = A_1 x + A_2\left(1 - e^{-a_1 x}\right) + A_3\left(1 - e^{-a_2 x}\right) - \int_0^x \left[A e^{-a_1(x-z)} + B e^{-a_2(x-z)}\right] \frac{dp'}{dz} dz;$$

the integrodifferential equation obtained here is

$$cp' = \varphi(x) - \int_0^x K(x, z) \frac{dp'}{dz} dz. \tag{12.9}$$

This equation is easily solved by operational methods. Subjecting both sides of (12.9) to the Laplace transformation with parameter s, we have

$$cP(s) = \Phi(s) - L(s) P'(s),$$

which we may rewrite on the basis of the property of differentiation of the transform [8] as

$$cP(s) = \Phi(s) - sL(s) P(s),$$

from which

$$P(s) = \frac{\Phi(s)}{c - sL(s)}, \tag{12.10}$$

where P(s), Φ(s), L(s) are the transforms of the functions p', φ(x), K(x), respectively. In this case

$$\Phi(s) = \int_0^\infty \varphi(x)e^{-sx}\,dx = \int_0^\infty \left[A_1 x + A_2(1 - e^{-\alpha_1 x}) + A_3(1 - e^{-\alpha_2 x}) \right] e^{-sx}\,dx =$$

$$= \frac{(A_1 + A_2\alpha_1 + A_3\alpha_2)\,s^2 + \left[A_1(\alpha_1 + \alpha_2) + (A_2 + A_3)\,\alpha_1\alpha_2 \right]s + A_1\alpha_1\alpha_2}{s^2(s + \alpha_1)(s + \alpha_2)},$$

$$L(s) = \int_0^\infty K(x)e^{-sx}\,dx = \int_0^\infty \left(Ae^{-\alpha_1 x} + Be^{-\alpha_2 x} \right) e^{-sx}\,dx = \frac{(A + B)s + A\alpha_2 + B\alpha_1}{(s + \alpha_1)(s + \alpha_2)}.$$

Then, according to (12.10),

$$P(s) = \frac{(A_1 + A_2\alpha_1 + A_3\alpha_2)\,s^2 + \left[A_1(\alpha_1 + \alpha_2) + (A_2 + A_3)\,\alpha_1\alpha_2 \right]s + A_1\alpha_1\alpha_2}{s^2\left\{ (A + B + c)\,s^2 + \left[A\alpha_2 + B\alpha_1 + c(\alpha_1 + \alpha_2) \right]s + c\alpha_1\alpha_2 \right\}}.$$

The original of this function is expressed as

$$p' = \frac{1}{2\pi i} \int_{\delta - i\infty}^{\delta + i\infty} P(s)e^{sx}\,ds = \frac{1}{2\pi i} \int_{\delta - i\infty}^{\delta + i\infty} \frac{(A_1A_2\alpha_1 + A_2\alpha_2)\,s^2 + \left[A_1(\alpha_1 + \alpha_2) + (A_2 + A_3)\,\alpha_1\alpha_2 \right]s + A_1\alpha_1\alpha_2}{s^2\left\{ (A + B + c)\,s^2 + \left[A\alpha_2 + B\alpha_1 + c(\alpha_1 + \alpha_2) \right]s + c\alpha_1\,\alpha_2 \right\}} e^{sx}\,ds.$$

Evaluating the last integral by utilizing residue theory [8], we obtain

$$p' = B_1 + B_2 x + B_3 e^{-\beta_1 x} + B_4 e^{-\beta_2 x}; \tag{12.11}$$

where

$$\left. \begin{aligned}
B_1 &= \frac{c(A_2 + A_3)\,\alpha_1\alpha_2 - A(A\alpha_2 + B\alpha_1)}{c^2\alpha_1\alpha_2} \\[6pt]
B_2 &= \frac{A_1}{c} \\[6pt]
B_3 &= \frac{(A_1 + A_2\alpha_1 + A_2\alpha_2)\,\beta_1^2 - \left[A_1(\alpha_1 + \alpha_2) + (A_2 + A_3)\,\alpha_1\alpha_2 \right]\beta_1 + A_1\alpha_1\alpha_2}{\beta_1^2\left\{ -2(A + B + c)\,\beta_1 + \left[A\alpha_2 + B\alpha_1 + c(\alpha_1 + \alpha_2) \right] \right\}} \\[6pt]
B_4 &= \frac{(A_1 + A_2\alpha_1 + A_3\alpha_2)\,\beta_2^2 - \left[A_1(\alpha_1 + \alpha_2) + (A_2 + A_3)\,\alpha_1\alpha_2 \right]\beta_2 + A_1\alpha_1\alpha_2}{\beta_2^2\left\{ -2(A + B + c)\,\beta_2 + \left[A\alpha_2 + B\alpha_1 + c(\alpha_1 + \alpha_2) \right] \right\}} \\[6pt]
\beta_1 &= \frac{A\alpha_2 + B\alpha_1 + c(\alpha_1 + \alpha_2)}{2(A + B + c)} - \sqrt{\left[\frac{A\alpha_2 + B\alpha_1 + c(\alpha_1 + \alpha_2)}{2(A + B + c)} \right]^2 - \frac{c\alpha_1\alpha_2}{A + B + c}} \\[6pt]
\beta_2 &= \frac{A\alpha_2 + B\alpha_1 + c(\alpha_1 + \alpha_2)}{2(A + B + c)} + \sqrt{\left[\frac{A\alpha_2 + B\alpha_1 + c(\alpha_1 + \alpha_2)}{2(A + B + c)} \right]^2 - \frac{c\alpha_1\alpha_2}{A + B + c}}
\end{aligned} \right\} \tag{12.12}$$

Finally, substituting the value of p' from (12.11) into (12.7), and evaluating the integrals, we have

$$u_1' = -\left(a_{11}B_1 + a_{15}\frac{B_2}{\alpha_1} + a_{16}\frac{B_2}{\alpha_2} - a_{13} - a_{14} \right) - (a_{11}B_2 - a_{12})\,x - a_{15}\left(\frac{\beta_1 B_3}{\alpha_1 - \beta_1} + \frac{\beta_2 B_4}{\alpha_1 - \beta_2} + \frac{a_{13}}{a_{15}} - \frac{B_2}{\alpha_1} \right)e^{-\alpha_1 x} -$$

$$- a_{16}\left(\frac{\beta_1 B_3}{\alpha_2 - \beta_1} + \frac{\beta_2 B_4}{\alpha_2 - \beta_2} - \frac{a_{14}}{a_{16}} - \frac{B_2}{\alpha_2} \right)e^{-\alpha_2 x} + B_3\left(a_{15}\frac{\beta_1}{\alpha_1 - \beta_1} + a_{16}\frac{\beta_1}{\alpha_2 - \beta_1} \right)e^{-\beta_1 x} + B_4\left(a_{15}\frac{\beta_2}{\alpha_1 - \beta_2} + a_{16}\frac{\beta_2}{\alpha_2 - \beta_2} \right)e^{-\beta_2 x},$$

$$u_2' = -\left(a_{11}B_1 + a_{25}\frac{B_2}{a_1} + a_{26}\frac{B_2}{a_2} - a_{23} - a_{24}\right) - (a_{11}B_2 - a_{12})x - a_{25}\left(\frac{\beta_1 B_3}{a_1 - \beta_1} + \frac{\beta_2 B_4}{a_1 - \beta_2} + \frac{a_{23}}{a_{25}} - \frac{B_2}{a_1}\right)e^{-a_1 x} -$$

$$- a_{26}\left(\frac{\beta_1 B_3}{a_2 - \beta_1} + \frac{\beta_2 B_4}{a_2 - \beta_2} + \frac{a_{24}}{a_{26}} - \frac{B_2}{a_2}\right)e^{-a_2 x} + B_3\left(a_{25}\frac{\beta_1}{a_1 - \beta_1} + a_{26}\frac{\beta_1}{a_2 - \beta_1}\right)\cdot e^{-\beta_1 x} + B_4\left(a_{25}\frac{\beta_2}{a_1 - \beta_2} + a_{26}\frac{\beta_2}{a_2 - \beta_2}\right)e^{-\beta_2 x},$$

$$u_3' = -\left(a_{11}B_1 + a_{35}\frac{B_2}{a_1} + a_{36}\frac{B_2}{a_2} - a_{33} - a_{34}\right) - (a_{11}B_2 + a_{12})x - a_{35}\left(\frac{\beta_1 B_3}{a_1 - \beta_1} + \frac{\beta_2 B_4}{a_1 - \beta_2} + \frac{a_{33}}{a_{35}} - \frac{B_2}{a_1}\right)e^{-a_1 x} -$$

$$- a_{36}\left(\frac{\beta_1 B_3}{a_2 - \beta_1} + \frac{\beta_2 B_4}{a_2 - \beta_2} + \frac{a_{34}}{a_{36}} - \frac{B_2}{a_2}\right)e^{-a_2 x} + B_3\left(a_{35}\frac{\beta_1}{a_1 - \beta_1} + a_{36}\frac{\beta_1}{a_2 - \beta_1}\right)e^{-\beta_1 x} + B_4\left(a_{35}\frac{\beta_2}{a_1 - \beta_2} + a_{36}\frac{\beta_2}{a_2 - \beta_2}\right)e^{-\beta_2 x}.$$

We determine ρ_1', ρ_2', ρ_3' by inserting the values of u_1', u_2', u_3' into (12.2).

Hence, the problem of steady motion of an n-component medium may be solved by the method of reduction to an integrodifferential equation. The only difficulty is that the characteristic equations will be of higher order; for example, we actually obtain a cubic algebraic equation to determine the roots λ_j in the case of a four-component medium; similar considerations apply to the evaluation of the residues of the functions.

CHAPTER III

STEADY MOTION OF VISCOUS AND VISCOUS-IDEAL MULTIPHASE MEDIA IN CONDUITS IN THE CONSTANT-POROSITY CASE

§ 13. Motion of Viscous Two-Phase Media between Two Parallel Walls

Let there be the motion of a two-phase medium between parallel walls extending to infinity on both sides. We direct the x axis along the center line between the walls, and consider the y axis to lie in the plane containing the x axis and perpendicular to the walls.

If the media of the two-phase mixture are incompressible, i.e.,

$$\rho_{1i} = \text{const}, \ \rho_{2i} = \text{const},$$

and the reduced densities are constant,

$$\rho_1 = \text{const}, \ \rho_2 = \text{const},$$

the porosities will also be constant:

$$f_1 = \text{const}, \ f_2 = \text{const}.$$

This means that we consider motion where the volume occupied by each of the media in unit volume of the two-phase mixture is taken to be the same at any point of the motion.

Let the two-phase mixture be in mutually penetrating steady, rectilinear motion parallel to the 0x [9]. Then

$$\frac{\partial u_1}{\partial t} = 0, \ \frac{\partial u_2}{\partial t} = 0$$

and moreover

$$v_1 = 0, \ v_2 = 0.$$

Under these conditions, the continuity equations (3.2) yield

$$\frac{\partial u_1}{\partial x} = 0, \ \frac{\partial u_2}{\partial x} = 0,$$

and we obtain from the second equation of (6.2) and (6.3)

$$\frac{\partial p}{\partial y} = 0.$$

35

The first equations of the system (6.2), (6.3) may therefore be written as

$$f_1\mu_1 \frac{\partial^2 u_1}{\partial y^2} + K(u_1 - u_2) = f_1\frac{\partial p}{\partial x} \Bigg\}$$
$$f_2\mu_2 \frac{\partial^2 u_2}{\partial y^2} + K(u_2 - u_1) = f_2\frac{\partial p}{\partial x} \Bigg\}$$

(13.1)

The left side in these equations is independent of x, but the right side depends on x. Such equations may exist if the right side is constant, i.e.,

$$\frac{\partial p}{\partial x} = N = \text{const.}$$

Then (13.1) is rewritten as

$$f_1\mu_1 \frac{d^2 u_1}{dy^2} + K(u_2 - u_1) = f_1 N$$

(13.2)

$$f_2\mu_2 \frac{d^2 u_2}{dy^2} + K(u_1 - u_2) = f_2 N.$$

(13.3)

If the spacing between the walls is 2h, the equations of these walls will be

$$y = \pm h.$$

To determine u_1 and u_2 we have the system of equations (13.2), (13.3), (2.1) and the boundary conditions

$$u_1 = u_2 = 0 \quad \text{for} \quad y = \pm h,$$

(13.4)

which result from the requirement that the media adhere to the fixed boundary walls. Adding (13.2) and (13.3) term by term, and taking account of (2.1), we have

$$f_1\mu_1 \frac{d^2 u_1}{dy^2} + f_2\mu_2 \frac{d^2 u_2}{dy^2} = N.$$

Furthermore, integrating twice with the boundary conditions, we obtain a relationship between the velocities of the media in the case under consideration:

$$f_1\mu_1 u_1 + f_2\mu_2 u_2 = -\frac{N}{2}(h^2 - y^2).$$

(13.5)

Solving (13.5) first with respect to u_2 and then with respect to u_1, and substituting the expressions obtained into (13.2) and (13.3), respectively, we obtain

$$\frac{d^2 u_1}{dy^2} - K\left(\frac{1}{f_1\mu_1} + \frac{1}{f_2\mu_2}\right)u_1 = -\frac{KN}{2f_1\mu_1 f_2\mu_2}(y^2 - h^2) + \frac{N}{\mu_1},$$

(13.6)

$$\frac{d^2 u_2}{dy^2} - K\left(\frac{1}{f_1\mu_1} + \frac{1}{f_2\mu_2}\right)u_2 = -\frac{KN}{2f_1\mu_1 f_2\mu_2}(y^2 - h^2) + \frac{N}{\mu_2},$$

(13.7)

Each of these equations is an inhomogeneous second-order linear equation with constant coefficients. The solutions of these equations under the boundary conditions (13.4) will be

$$u_1 = -\frac{\partial p}{\partial x}\left[A_1 \text{ch} my + By^2 + C_1\right],$$

(13.8)

$$u_2 = -\frac{\partial p}{\partial x}\left[A_2 \text{ch} my + By^2 + C_2\right],$$

(13.9)

where

$$A_1 = \left(\frac{1}{f_1\mu_1 + f_2\mu_2} - \frac{1}{\mu_1} \right) \frac{1}{m^2 \text{ch} mh}$$

$$A_2 = \left(\frac{1}{f_1\mu_1 + f_2\mu_2} - \frac{1}{\mu_2} \right) \frac{1}{m^2 \text{ch} mh}$$

$$B = - \frac{1}{2(f_1\mu_1 + f_2\mu_2)}$$

$$C_1 = -\frac{1}{m^2} \left(\frac{1}{f_1\mu_1 + f_2\mu_2} - \frac{1}{\mu_1} - \frac{Kh^2}{2 f_1\mu_1 f_2\mu_2} \right)$$

$$C_2 = -\frac{1}{m^2} \left(\frac{1}{f_1\mu_1 + f_2\mu_2} - \frac{1}{\mu_2} - \frac{Kh^2}{2 f_1\mu_1 f_2\mu_2} \right)$$

$$m^2 = K \frac{f_1\mu_1 + f_2\mu_2}{f_1\mu_1 f_2\mu_2}$$

$$(13.10)$$

Formulas (13.8) and (13.9) give the distributions of the velocities u_1 and u_2 over the cross section. As is seen, the velocities are directly proportional to the pressure drop per unit length.

Let us determine the signs of the notation in (13.10).

Let us perform the following transformation:

$$f_1\mu_1 + f_2\mu_2 = (1 - f_2)\mu_1 + f_2\mu_2 = \mu_1 + f_2(\mu_2 - \mu_1),$$

analogously

$$f_1\mu_1 + f_2\mu_2 = \mu_2 - f_1(\mu_2 - \mu_1).$$

For concreteness let us put $\mu_2 > \mu_1$; we then obtain the following inequality from the two preceding equalities:

$$\mu_2 > f_1\mu_1 + f_2\mu_2 > \mu_1,$$

from which

$$\frac{1}{f_1\mu_1 + f_2\mu_2} - \frac{1}{\mu_1} < 0,$$

$$\frac{1}{f_1\mu_1 + f_2\mu_2} - \frac{1}{\mu_2} > 0,$$

$$\frac{1}{f_1\mu_1 + f_2\mu_2} - \frac{1}{\mu_1} - \frac{Kh^2}{2 f_1\mu_1 f_2\mu_2} < 0.$$

Moreover, from (13.9) we have the condition $A_2 + C_2 > 0$, under which $u_2 > 0$. Therefore

$$\left. \begin{array}{l} A_1 < 0 \\ B_1 < 0 \\ C_1 > 0 \\ A_2 > 0 \\ C_2 > 0 \quad \text{or} \quad C_2 < 0 \end{array} \right\}$$

$$(13.11)$$

Utilizing (13.8) and (13.9), we write for the difference in velocities of the media

$$u_1 - u_2 = -\frac{\partial p}{\partial x}\frac{1}{m^2}\left[\left(\frac{1}{\mu_1} - \frac{1}{\mu_2}\right)\frac{\mathrm{ch}\,mh - \mathrm{ch}\,my}{\mathrm{ch}\,mh}\right] \geqslant 0. \tag{13.12}$$

This difference between the velocities is greater than zero since y may vary between the limits –h ≤ y ≤ h, for which

$$\mathrm{ch}\,mh - \mathrm{ch}\,my \geqslant 0.$$

Moreover

$$\frac{1}{\mu_1} - \frac{1}{\mu_2} > 0$$

(because of the assumption that $\mu_2 > \mu_1$).

Starting from (13.12), the result obtained may be formulated as follows: **The less viscous of two media moving in combination has the greater velocity.**

The general form of the velocity distribution curves is given in Fig. 3. These curves are not parabolas, as in the motion of an incompressible single-phase fluid, but are curves described by the sum of the hyperbolic cosine and a quadratic binomial.

The difference $u_1 - u_2$ depends on the magnitude of the difference $\mu_2 - \mu_1$ and increases with increase in the latter.

It is easy to note that for the same viscosities in the media, i.e., for

$$\mu_1 = \mu_2 = \mu,$$

we have $u_1 = u_2 = u$ from (13.8) and (13.9). In this case the graphs of the velocity distributions of both media coincide and yield a parabola, and the formula of the velocity distribution of a single-phase incompressible fluid over a cross section becomes [10]

$$u = -\frac{1}{2\mu}\frac{\partial p}{\partial x}(h^2 - y^2). \tag{13.13}$$

Let us find the change in the relative velocity of the media as a percentage of the velocity of the first medium along the axis of a plane conduit.

It is easy to see that by using the velocity formulas (13.8) and (13.9) in dimensionless parameters, this change will be

$$\frac{u_{1max} - u_{2max}}{u_{1max}}\cdot 100\% = \cfrac{\mu - 1}{\mu - \cfrac{\mu}{f_1+f_2\mu} + \cfrac{\frac{1}{2f_1f_2}H\mathrm{ch}\sqrt{\frac{\frac{f_1}{\mu}+f_2}{f_1f_2}H}}{\mathrm{ch}\sqrt{\frac{\frac{f_1}{\mu}+f_2}{f_1f_2}H}-1}}\times 100\%, \tag{13.14}$$

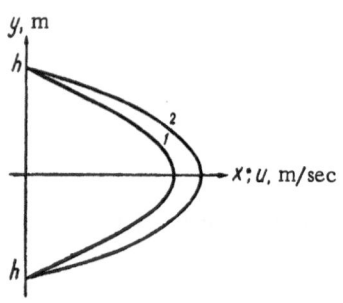

Fig. 3. Velocity distribution curve. 1) More viscous phase; 2) less viscous phase.

where $\mu = \mu_2/\mu_1$ and $H = Kh^2/\mu_1$. u_{1max} and u_{2max} are respectively, the values of u_1 and u_2 at y = 0. As is seen, (13.14) depends on the dimensionless quantities μ, H, f_1, or f_2. Computations have shown that the relative velocity of the media in percentages of the velocity of the first medium increases as the volume contained by the first medium decreases.

Analogous investigations can be made for each of the dimensionless parameters. However, it is here necessary to take account of the experimental values of the other parameters. This becomes clear if we consider that H changes as μ changes.

Let us now find the volume discharge of the media. Let Q_1 and Q_2 denote the volume discharges of the media; then

$$Q_1 = f_1 \int_{-h}^{h} u_1 dy,$$

$$Q_2 = f_2 \int_{-h}^{h} u_2 dy.$$

Substituting the expressions for u_1 and u_2 from (13.8) and (13.9) into the preceding formulas and integrating, we obtain

$$Q_1 = -\frac{\partial p}{\partial x} f_1 \left\{ 2 \left(\frac{1}{f_1\mu_1 + f_2\mu_2} - \frac{1}{\mu_1} \right) \left[\frac{f_1\mu_1 f_2\mu_2}{K(f_1\mu_1 + f_2\mu_2)} \right]^{\frac{3}{2}} \times \right.$$

$$\left. \times \text{th} \sqrt{K \frac{f_1\mu_1 + f_2\mu_2}{f_1\mu_1 f_2\mu_2}} \; h + \frac{2h^3}{3(f_1\mu_1 + f_2\mu_2)} - \left(\frac{1}{f_1\mu_1 + f_2\mu_2} - \frac{1}{\mu_1} \right) \frac{2 f_1\mu_1 f_2\mu_2}{K(f_1\mu_1 + f_2\mu_2)} \; h \right\}, \qquad (13.15)$$

$$Q_2 = -\frac{\partial p}{\partial x} f_2 \left\{ 2 \left(\frac{1}{f_1\mu_1 + f_2\mu_2} - \frac{1}{\mu_2} \right) \left[\frac{f_1\mu_1 f_2\mu_2}{K(f_1\mu_1 + f_2\mu_2)} \right]^{\frac{3}{2}} \text{th} \sqrt{K \frac{f_1\mu_1 + f_2\mu_2}{f_1\mu_1 f_2\mu_2}} \; h + \right.$$

$$\left. + \frac{2h^3}{3(f_1\mu_1 + f_2\mu_2)} - \left(\frac{1}{f_1\mu_1 + f_2\mu_2} \right) \frac{2 f_1\mu_1 f_2\mu_2}{K(f_1\mu_1 + f_2\mu_2)} \cdot h \right\}. \qquad (13.16)$$

It thus follows from (13.15) and (13.16) that the volume discharge of each of the media is proportional to the pressure drop per unit length, and if $\mu_1 = \mu_2 = \mu$ then the total volume discharge will be

$$Q = Q_1 + Q_2 = -\frac{\partial p}{\partial x} \frac{2h^3}{3\mu}.$$

This is the well-known Poiseuille formula for an incompressible viscous fluid [10].

§ 14. Motion of Viscous Two-Phase Media in a Circular Cylindrical Conduit

Let us consider the motion of a two-phase medium in a circular cylindrical conduit [9]. We shall start from an analysis of the differential equations of motion of viscous two-phase media in cylindrical coordinates.

Let the media constituting the mixture be incompressible and let the reduced densities be constant. Let us assume that the motion is rectilinear and parallel to the 0x axis (the cylinder axis); then

$$v_{1r} = v_{2r} = 0, \quad v_{1\varphi} = v_{2\varphi} = 0.$$

From the continuity equation we have

$$\frac{\partial u_1}{\partial x} = 0, \quad \frac{\partial u_2}{\partial x} = 0.$$

The equations of motion in cylindrical coordinates (8.3) become

$$f_1\mu_1 \left(\frac{\partial^2 u_1}{\partial r^2} + \frac{1}{r^2} \frac{\partial^2 u_1}{\partial \varphi^2} + \frac{1}{r} \frac{\partial u_1}{\partial r} \right) + K (u_2 - u_1) = f_1 \frac{\partial p}{\partial x}$$

$$f_2\mu_2 \left(\frac{\partial^2 u_2}{\partial r^2} + \frac{1}{r^2} \frac{\partial^2 u_2}{\partial \varphi^2} + \frac{1}{r} \frac{\partial u_2}{\partial r} \right) + K (u_1 + u_2) = f_2 \frac{\partial p}{\partial x}$$

$$(14.1)$$

By reasoning exactly as in § 13, we establish that

$$\frac{\partial p}{\partial x} = N = \text{const}$$

for the pressure.

Let us consider the case when the steady rectilinearly-parallel flow of incompressible viscous media participating in the motion of a two-phase mixture in a circular cylindrical conduit possesses axial symmetry, i.e., when

$$\frac{\partial u_1}{\partial \varphi} = 0, \qquad \frac{\partial u_2}{\partial \varphi} = 0.$$

Then (14.1) is rewritten as

$$f_1\mu_1 \left(\frac{d^2 u_1}{dr^2} + \frac{1}{r} \frac{du_1}{dr} \right) + K (u_2 - u_1) = f_1 N, \tag{14.2}$$

$$f_2\mu_2 \left(\frac{d^2 u_2}{dr^2} + \frac{1}{r} \frac{du_2}{dr} \right) + K (u_1 - u_2) = f_2 N. \tag{14.3}$$

To determine u_1 and u_2 we have (14.2), (14.3), (2.1), and the boundary conditions

$$\left. \begin{array}{l} u_1 = 0 \\ u_2 = 0 \end{array} \right\} \tag{14.4}$$

where $r = R$ (R is the radius of the conduit), which follow from the requirement that the media adhere to the conduit walls.

Adding (14.2) and (14.3) term by term, and also taking account of (2.1), we obtain the expression

$$f_1\mu_1 \frac{d}{dr} \left(r \frac{du_1}{dr} \right) + f_2\mu_2 \frac{d}{dr} \left(r \frac{du_2}{dr} \right) = Nr,$$

which after double integration yields

$$f_1\mu_1 u_1 + f_2\mu_2 u_2 = \frac{N}{4} r^2 + c_1 \ln r + c_2. \tag{14.5}$$

However $\ln r$ becomes infinite for $r = 0$. To avoid this we put $c_1 = 0$. We determine the unknown constant c_2 from boundary condition (14.4), after which (14.5) becomes

$$f_1\mu_1 u_1 + f_2\mu_2 u_2 = -\frac{N}{4} (R^2 - r^2). \tag{14.6}$$

Solving (14.6) for u_2 and u_1 and substituting the solutions in (14.2) and (14.3), we obtain the equations

$$\frac{d^2 u_1}{dr^2} + \frac{1}{r} \frac{du_1}{dr} - m^2 u_1 = Ar^2 + B, \tag{14.7}$$

$$\frac{d^2 u_2}{dr^2} + \frac{1}{r} \frac{du_2}{dr} - m^2 u_2 = Ar^2 + D, \tag{14.8}$$

where

$$A = -\frac{KN}{4f_1\mu_1 f_2\mu_2}\ , \quad B = \frac{KNR^2}{4f_1\mu_1 f_2\mu_2} + \frac{N}{\mu_1}$$
$$D = \frac{KNR^2}{4f_1\mu_1 f_2\mu_2} + \frac{N}{\mu_2}, \quad m^2 = K\left[\frac{1}{f_1\mu_1} + \frac{1}{f_2\mu_2}\right]$$

(14.9)

As is seen, (14.7) and (14.8) differ only in the given constants B and D, and the boundary conditions for u_1 and u_2 are identical. Hence, it is sufficient to solve (14.7) and then by replacing B by D to obtain the solution of (14.8).

Let us substitute ξ = mr into (14.7), which then will become

$$\frac{d^2u_1}{d\xi^2} + \frac{1}{\xi}\frac{du_1}{d\xi} - u_1 = \frac{A}{m^4}\xi^2 + \frac{B}{m^2}.$$

(14.10)

We solve (14.10) by the Lagrange method. The homogeneous differential equation, the Bessel equation in our case

$$\frac{d^2u_1}{d\xi^2} + \frac{1}{\xi}\frac{du_1}{d\xi} - u_1 = 0,$$

(14.11)

has the solution

$$u_1^* = c_1 I_0(\xi) + c_2 K_0(\xi),$$

(14.12)

where $I_0(\xi)$ is the Bessel function of imaginary argument, and $K_0(\xi)$ is the Macdonald function.

Considering c_1 and c_2 as variables, we may write

$$c_1' I_0(\xi) + c_2' K_0(\xi) = 0,$$

$$c_1' I_0(\xi) + c_2' K_0'(\xi) = \frac{A}{m^4}(\xi^2) + \frac{B}{m^2},$$

from which

$$c_2 = \frac{1}{m^2}\int \frac{\left(\frac{A}{m^2}\xi^2 + B\right)I_0(\xi)\,d\xi}{I_0'(\xi)K_0(\xi) - K_0'(\xi)I_0(\xi)}.$$

On the basis of the known equality

$$I_0'(\xi)K_0(\xi) - K_0'(\xi)I_0(\xi) = \frac{1}{\xi}$$

(14.13)

we have

$$c_1 = \frac{1}{m^2}\int\left(\frac{A}{m^2}\xi^3 + B\xi\right)K_0(\xi)\,d\xi$$
$$c_2 = \frac{1}{m^2}\int\left(\frac{A}{m^2}\xi^3 + B\xi\right)I_0(\xi)\,d\xi$$

(14.14)

Utilizing the formulas [11]

$$I_0'(\xi) = I_1(\xi)$$
$$K_0'(\xi) = -K_1(\xi)$$

(14.15)

$$\xi K_1'(\xi) + K_1(\xi) = - \xi K_0(\xi)$$
$$\xi I_1'(\xi) + I_1(\xi) = \xi I_0(\xi)$$

we evaluate the integrals

$$\int \xi K_0(\xi)\, d\xi = - \xi K_1(\xi) + c_{01}$$
$$\int \xi I_0(\xi)\, d\xi = \xi I_1(\xi) + c_{02}$$
$$\int \xi^2 K_1(\xi)\, d\xi = - \xi^2 K_0(\xi) - 2\xi K_1(\xi) + c_{03}$$
$$\int \xi^3 K_0(\xi)\, d\xi = - \xi^3 K_1(\xi) - 2\xi^2 K_0(\xi) - 4\xi K_1(\xi) + c_{04}$$
$$\int \xi^2 I_1(\xi)\, d\xi = \xi^2 I_0(\xi) - 2\xi I_1(\xi) + c_{05}$$
$$\int \xi^3 I_1(\xi)\, d\xi = \xi^3 I_1(\xi) - 2\xi^2 I_0(\xi) + 4\xi I_1(\xi) + c_{06}$$

(14.16)

On the basis of (14.16), the expression (14.14) may be represented as

$$c_1 = - \frac{A}{m^4}\, \xi^3 K_1(\xi) - \frac{2A}{m^4}\, \xi^2 K_0(\xi) - \left(\frac{4A}{m^4} + \frac{B}{m^2} \right) \xi K_1(\xi) + c_3$$
$$c_2 = - \frac{A}{m^4}\, \xi^3 I_1(\xi) + \frac{2A}{m^4}\, \xi^2 I_0(\xi) - \left(\frac{4A}{m^4} + \frac{B}{m^2} \right) \xi I_1(\xi) + c_4$$

(14.17)

Substituting (14.17) into (14.12), utilizing (14.13), and replacing ξ by mr, we have

$$u_1 = c_5 I_0(mr) + c_6 K(mr) - \frac{A}{m^2}\, r^2 - \frac{4A}{m^4} - \frac{B}{m^2}.$$

Since $K_0(mr)$ becomes infinite for r = 0, we then put $c_6 = 0$; then after c_5 has been determined by utilizing the boundary condition (14.4), the preceding formula becomes

$$u_1 = \left[\frac{A}{m^2} R^2 + \frac{4A}{m^4} + \frac{B}{m^2} \right] \frac{I_0(mr)}{I_0(mR)} - \frac{A}{m^2}\, r^2 - \frac{4A}{m^4} - \frac{B}{m^2}.$$

(14.18)

Replacing B by D, we obtain for u_2

$$u_2 = \left[\frac{A}{m^2} R^2 + \frac{4A}{m^4} + \frac{D}{m^2} \right] \frac{I_0(mr)}{I_0(mR)} - \frac{A}{m^2}\, r^2 - \frac{4A}{m^4} - \frac{D}{m^2}.$$

(14.19)

Substituting A, B, and D from (14.9) into (14.18) and (14.19), we have

$$u_1 = - \frac{1}{m^2} \frac{\partial p}{\partial x} \left[\left(\frac{1}{f_1\mu_1 + f_2\mu_2} - \frac{1}{\mu_1} \right) \frac{I_0(mr)}{I_0(mR)} + \frac{K}{4 f_1\mu_1 f_2\mu_2} (R^2 - r^2) - \left(\frac{1}{f_1\mu_1 + f_2\mu_2} - \frac{1}{\mu_1} \right) \right],$$

(14.20)

$$u_2 = - \frac{1}{m^2} \frac{\partial p}{\partial x} \left[\left(\frac{1}{f_1\mu_1 + f_2\mu_2} - \frac{1}{\mu_2} \right) \frac{I_0(mr)}{I_0(mR)} + \frac{K}{4 f_1\mu_1 f_2\mu_2} (R^2 - r^2) - \left(\frac{1}{f_1\mu_1 + f_2\mu_2} - \frac{1}{\mu_2} \right) \right].$$

(14.21)

Formulas (14.20) and (14.21) determine the velocity distribution over the cross section of a circular cylindrical conduit of each of the media in the moving mixture.

As is seen, in the case under consideration the velocity of each medium is distributed over the cross section of the circular cylindrical conduit along a curve mapped by the sum of two functions; the zero-order Bessel function of imaginary argument and a quadratic binomial.

When the viscosities of all the media are identical, the velocities are all equal, and the velocity distribution curve will be a parabola as in the case of a single-phase incompressible fluid:

$$u = -\frac{1}{4\mu} \frac{\partial p}{\partial x} (R^2 - r^2).$$

The changes in motion parameters in a circular cylindrical conduit agree qualitatively with the changes in the same parameters in motion in a plane conduit (see §13). We determine the volume discharge by means of the formulas

$$Q_1 = f_1 \int_0^R 2\pi r u_1 dr,$$

$$Q_2 = f_2 \int_0^R 2\pi r u_2 dr.$$

Substituting (14.20) and (14.21) for u_1 and u_2 in these formulas, and integrating between 0 and R, we obtain

$$Q_1 = -\frac{2\pi r f_1}{m^2} \frac{\partial p}{\partial x} \left[\left(\frac{1}{f_1\mu_1 + f_2\mu_2} - \frac{1}{\mu_1} \right) \frac{I_1(mR)}{mI_0(mR)} + \left(\frac{KR^2}{8f_1\mu_1 f_2\mu_2} + \frac{1}{\mu_1} - \frac{1}{f_1\mu_1 + f_2\mu_2} \right) \frac{R}{2} \right], \qquad (14.22)$$

$$Q_2 = -\frac{2\pi R f_2}{m^2} \frac{\partial p}{\partial x} \left[\left(\frac{1}{f_1\mu_1 + f_2\mu_2} - \frac{1}{\mu_2} \right) \frac{I_1(mR)}{mI_0(mR)} + \left(\frac{KR^2}{8f_1\mu_1 f_2\mu_2} + \frac{1}{\mu_2} - \frac{1}{f_1\mu_1 + f_2\mu_2} \right) \frac{R}{2} \right]. \qquad (14.23)$$

Thus, in this case also the volume discharge is directly proportional to the pressure drop per unit length.

§15. Stationary Flow of a Conducting Medium in a Magnetic Field

Let us consider the problem of the motion of a two-phase medium, one of whose phases is conducting [12]. Such a mixture may evidently be described by the system of equations given in Chapter I if an additional electromagnetic transmission term is included in the equation of motion for the conducting medium and this system is supplemented by the Maxwell equations. A study of such a system may be used to determine, for example, how a nonconducting fluid affects the motion of a conducting fluid.

Limiting ourselves to the consideration of two incompressible media, and retaining the notation of magnetohydrodynamics, we may write

$$\left. \begin{aligned} \operatorname{div} \vec{V} &= 0 \\ \rho \frac{d\vec{V}}{dt} &= -f \frac{\partial p}{\partial x} + K\left(\vec{V}_n - \vec{V}\right) + \mu\left[\vec{j}\vec{H}\right] + \eta_1 f \Delta \vec{V} \\ \operatorname{div} \vec{V}_n &= 0 \\ \rho_n \frac{d\vec{V}_n}{dt} &= -f_n \frac{\partial p}{\partial x} + K\left(\vec{V} - \vec{V}_n\right) + \eta_n f_n \Delta \vec{V}_n \\ f + f_n &= 1 \end{aligned} \right\} \qquad (15.1)$$

where the first two equations refer to the conducting fluid, the second two to the nonconducting fluid (the corresponding quantities are denoted with a subscript n); η is the dynamic coefficient of viscosity; f, f_n the porosities; $\vec{j}$ the electric current density; $\vec{H}$ the magnetic field intensity; and μ the magnetic permittivity of the medium.

The system (15.1) should be supplemented with the Maxwell equations.

As an illustration of the applicability of the system (15.1), let us examine the problem of stationary one-dimensional flow of a conducting fluid mixed with a nonconducting fluid between two parallel planes in the presence of a transverse external magnetic field. Let the motion of the media be in the x direction; let the external uniform magnetic field $\vec{H_0}$ be directed along z; and let the boundary planes be $z = \pm l$.

As is known, the flow stretches the lines of force of the perpendicular magnetic field. Hence, besides the transverse component H_0, a component h_x parallel to the motion is also manifest. This latter is a function of z since the velocity component along x depends on z. Without complicating the solution of the problem, we shall consider an external uniform electric field

$$E_y = E_0$$

to be applied in the 0y direction.

Taking account of the geometry of the problem in the stationary case, we obtain the following system to determine $v_x = v(z)$, $v_{nx} = v_n(z)$, $h_x = h(z)$:

$$\left.\begin{array}{c} f\eta \dfrac{d^2v}{dz^2} + \mu j_y H_0 - f\dfrac{\partial p}{\partial x} + K(v_n - v) = 0 \\[2mm] f_n \eta_n \dfrac{d^2 v_n}{dz^2} - f_n \dfrac{\partial p}{\partial x} + K(v - v_n) = 0 \\[2mm] \dfrac{dh}{dz} = 4\pi j_y \\[2mm] j_y = \sigma(E_0 - \mu H_0 v) \end{array}\right\} \qquad (15.2)$$

where σ is the conductivity of the medium and the pressure gradient is $\partial p/\partial x = \text{const}$. The third equation in the system (15.2) is the Maxwell equation, and the fourth is a relationship for the electric current density.

Utilizing the expression for j_y, we can determine v, v_n from the first two equations in the system (15.2). The problem is solved under the boundary conditions $v = v_n = 0$ for $z = \pm l$. The solution is

$$v = -\frac{1}{\tau_i}\frac{\partial p}{\partial x}\left(a\frac{\mathrm{ch}\,z/r_1}{\mathrm{ch}\,l/r_1} + b\frac{\mathrm{ch}\,z/r_2}{\mathrm{ch}\,l/r_2} + \frac{l^2}{M^2}\right) + u, \qquad (15.3)$$

$$v_n = -\frac{1}{\tau_{in}}\frac{\partial p}{\partial x}\left(a_n\frac{\mathrm{ch}\,z/r_1}{\mathrm{ch}\,l/r_1} + b_n\frac{\mathrm{ch}\,z/r_2}{\mathrm{ch}\,l/r_2} + \frac{l^2}{M_n^2} + \frac{l^2}{L_n^2}\right) + u_n, \qquad (15.4)$$

where u, u_n are the velocities due to the presence of the electric field E_0 perpendicular to H_0:

$$u = \frac{E_0}{\mu H_0}\left(a\frac{\mathrm{ch}\,z/r_1}{\mathrm{ch}\,l/r_1} + \beta\frac{\mathrm{ch}\,z/r_2}{\mathrm{ch}\,l/r_2} + 1\right); \qquad (15.5)$$

$$u_n = \frac{E_0}{\mu H_0}\left(a_n\frac{\mathrm{ch}\,z/r_1}{\mathrm{ch}\,l/r_1} + \beta_n\frac{\mathrm{ch}\,z/r_2}{\mathrm{ch}\,l/r_2} + 1\right). \qquad (15.6)$$

The following notation has been introduced in (15.3) - (15.6):

$$a = \delta\left[\left(\frac{l}{r_2}\right)^2 \frac{1}{M^2} - 1\right], \quad a_n = \delta\left[\left(\frac{l}{r_2}\right)^2\left(\frac{1}{M_n^2} + \frac{1}{L_n^2}\right) - 1\right],$$

$$b = -\delta\left[\left(\frac{l}{r_1}\right)^2 \frac{1}{M^2} - 1\right], \quad b_n = -\delta\left[\left(\frac{l}{r_1}\right)^2\left(\frac{1}{M_n^2} + \frac{1}{L_n^2}\right) - 1\right],$$

$$\alpha = \delta\left(\frac{1}{r_2^2} - \frac{M^2}{f}\frac{1}{l^2}\right), \quad \beta = -\delta\left(\frac{1}{r_1^2} - \frac{M^2}{f}\frac{1}{l^2}\right),$$

$$\alpha_n = \delta/r_2^2, \quad \beta_n = -\delta/r_1^2,$$

$$\delta = \frac{1}{\sqrt{s^2 - 4q}},$$

$$r_1 = \sqrt{\frac{2}{s + \sqrt{s^2 + 4q}}}, \quad r_2 = \sqrt{\frac{2}{s - \sqrt{s^2 - 4q}}}, \qquad (*)$$

$$s = 1/l^2\left(L^2 + L_n^2 + \frac{1}{f}M^2\right), \quad q = 1/l^4\frac{M^2 L_n^2}{f}, \qquad (**)$$

$$L = l\left(\frac{K}{f\eta}\right)^{1/2}, \quad L_n = l\left(\frac{K}{f_n\,\eta_n}\right)^{1/2},$$

$$M = \mu H_0 l\left(\frac{\sigma}{\eta}\right)^{1/2},$$

$$M_n = \mu H_0 l\left(\frac{\sigma}{\eta_n}\right)^{1/2};$$

where M is the Hartman number.

We have the following equation to determine the magnetic fluid h:

$$\frac{d^2 h}{dz^2} = -4\pi\sigma\mu H_0 \frac{dv}{dz}. \tag{15.7}$$

Substituting (15.3) here for v and integrating under the boundary conditions h = 0 when $z = \pm l$ (these conditions follow from the continuity of the tangential component of the magnetic field $\overrightarrow{H}$ to the boundary), we find

$$h = A\frac{\text{sh } z/r_1 - z/l \text{ sh } l/r_1}{\text{ch } l/r_1} + B\frac{\text{sh } z/r_2 - z/l \text{ sh } l/r_2}{\text{ch } l/r_2}, \tag{15.8}$$

where

$$A = 4\pi\sigma r_1\left(\frac{\mu H_0}{\eta}a\frac{\partial p}{\partial x} - \alpha E_0\right),$$

$$B = 4\pi\sigma r_2\left(\frac{\mu H_0}{\eta}b\frac{\partial p}{\partial x} - \beta E_0\right).$$

The mean value of the magnetic field h is zero for $-l < z < l$. For $E_0 = 0$ in the domain $sl^2 \ll 1$ it is proportional to H_0 and independent of the coefficient of interaction and the properties of the nonconducting fluid:

$$h = -\frac{2\pi}{3}\frac{\sigma\mu H_0}{\eta}\frac{\partial p}{\partial x}z(z^2 - l^2). \tag{15.9}$$

Expressions (15.3) - (15.4) yield the velocity distribution over the cross section; (15.8) and (15.9) determine the field distribution. The quantities r_1 and r_2 here are real since, as is easy to see from (*) and (**),

$$s^2 - 4q \geqslant 0.$$

The equality sign is valid for $H_0 = K = 0$.

In the limiting case of no magnetic and electric fields, the dependence of v and v_n on z should be the same as if both media were nonconducting. Passing to the limit in (15.3) − (15.4) for the case of motion of two-phase nonconducting media, we obtain the same result as in § 13. For $K \to 0$ and $f = 1$ formula (15.3) goes over into the well-known expression for the case of flow of a conducting fluid in a magnetic field [13].

The existence of both the magnetic field and the nonconducting medium is felt in the appearance of an additional resistance of the conducting fluid to the motion. The other parameters remaining the same, an increase in the field H_0 or the friction coefficient K results in a decrease in the velocities v and v_n. By changing the motion of the conducting medium, the magnetic field also affects the nonconducting fluid (because of friction).

We cannot assert that in the joint motion of two media in a magnetic field the medium with the lower viscosity has the greater velocity (see § 13) since the inequalities $v_n > v$ and $v_n < v$ are valid in this case. It may only be said that both for $\eta_n > \eta$, and $\eta_n < \eta$ the inequality

$$v_n - v \leqslant - \frac{f_n}{K} \frac{\partial p}{\partial x} \tag{15.10}$$

is satisfied (the equality sign refers to the case $\eta = \eta_n = 0$). In the domain where the viscosity η is small and the field H_0, is not too great, $v > v_n$ should be expected.

The velocity v_n also depends on the electric field. The presence of an electric field results in the appearance of an additional velocity u_n, where, as should be expected, $u_n < u$. In addition to the viscosity and induced deceleration, the dynamic friction between the mixture components also affects the velocity profile.

When $l \ll r_1$, $(sl^2 \ll 1)$, the influence of the magnetic field and the friction between the fluids on the motion becomes insignificant compared with internal friction in each component (η, η_n), and the expressions for the velocities have the customary parabolic form:

$$v = \frac{1}{2\eta} \left(\frac{\sigma E_0 H_0}{f} - \frac{\partial p}{\partial x} \right) (e^2 - z^2), \tag{15.11}$$

$$v_n = - \frac{1}{2\eta_n} \frac{\partial p}{\partial x} (l^2 - z^2). \tag{15.12}$$

When the term with the viscosity may be neglected in (15.2) (small viscosity, large magnetic field), we obtain (in the absence of a field E_0)

$$v_n = - \frac{l^2}{\eta_n} \frac{\partial p}{\partial x} \left(\frac{1}{M_n^2} + \frac{1}{L_n^2} \right) \left(1 - \frac{\operatorname{ch} z/r}{\operatorname{ch} l/r} \right), \tag{15.13}$$

$$v - v_n = - \frac{1}{L_2 + \dfrac{M^2}{f}} \frac{\partial p}{\partial x} \left[\frac{l^2}{\eta} - \frac{1}{K} \left(L^2 + \frac{f_n}{f} M^2 \right) \left(1 - \frac{\operatorname{ch} z/r}{\operatorname{ch} l/r} \right) \right], \tag{15.14}$$

where

$$r = l \sqrt{ \frac{f_n}{M_n^2} + \frac{1}{L_n^2} }.$$

Now the magnetic "viscosity" exerts the fundamental influence on the motion by hindering the passage of the mixture perpendicular to the lines of force.

For the domain $l \ll r$ the expressions (15.3) – (15.4) become

$$v_n = -\frac{1}{2\eta} \frac{\partial p}{\partial x} \frac{M_n^2 + L_n^2}{M_n^2 + f_n L_n^2} (l^2 - z^2),$$ (15.15)

$$v - v_n = -\frac{1}{2\eta_n} \frac{\partial p}{\partial x} \left[\frac{2fl^2}{L_n^2 + M_n^2} - \frac{M_n^2 (L_n^2 + M_n^2)}{(f_n L_n^2 + M_n^2)} (l^2 - z^2) \right].$$ (15.16)

Moreover, if the term with the viscosity η_n can be neglected in (15.2), we obtain expressions independent of z:

$$v = -\frac{1}{\sigma \mu^2 H_0^2} \frac{\partial p}{\partial x} + \frac{E_0}{\mu H_0},$$ (15.17)

$$v_n - v = \frac{f_n}{K} \frac{\partial p}{\partial x}.$$ (15.18)

It is seen from (15.17) that the velocity v decreases as $1/H_0^2$ as the field grows, and that v is independent of the coefficient of interaction. For very high values of K the velocities of both components are equalized. It should be noted that (15.17) and (15.18) are actually valid only far from the walls, where the viscosity cannot be substantial. However, its influence is always felt in the thin layer near the wall.

It is important to know the mean velocity in questions associated with the discharge of fluids (or gases). From (15.3) and (15.4) we have

$$\bar{v} = -\frac{1}{l} \frac{\partial p}{\partial x} \left(ar_1 \, \text{th} \, l/r_1 + br_2 \, \text{th} \, l/r_2 + \frac{l}{\sigma H_0^2} \right) + \bar{u},$$ (15.19)

$$\bar{v}_n = -\frac{1}{l} \frac{\partial p}{\partial x} \left(a_n r_1 \, \text{th} \, l/r_1 + b_n r_2 \, \text{th} \, l/r_2 + \frac{l}{\sigma H_0^2} + \frac{lf_n}{K} \right) + \bar{u}_n.$$ (15.20)

For domains where $1/r^2$ is of the order of several units or greater, (15.19) and (15.20) may be approximated by the expressions (we do not write terms proportional to E_0)

$$v = -\frac{1}{\eta} \frac{\partial p}{\partial x} \left\{ Q \left[\left(\frac{L}{M} \right)^2 + \left(\frac{L_n}{M} \right)^2 + \frac{1}{f^{1/2}} \frac{L_n}{M} + \frac{f_n}{f} \right] + \frac{l^2}{M^2} \right\},$$ (15.21)

$$v_n - v = -\frac{1}{\eta_n} \frac{\partial p}{\partial x} \left[Q \left(\frac{1}{f} \frac{M^2}{L_n^2} + \frac{1}{f^{1/2}} \frac{M}{L_n} + \frac{1}{f} \frac{\eta_n}{\eta} \right) + \frac{l^2}{L_n^2} \right],$$ (15.22)

where

$$Q = \frac{1}{l} \left[\sqrt{\frac{s - \sqrt{s^2 - 4q}}{2q(s^2 - 4q)}} - \sqrt{\frac{s + \sqrt{s^2 - 4q}}{2q(s^2 - 4q)}} \right].$$

Our results are valid for the case of motion between two parallel plane walls. However, they may be applied also to rectangular conduits used in experiment if one of the sides of the rectangle is much greater than the other and the magnetic field is perpendicular to the long side. The mean velocities v, v_n may be measured experimentally, which will afford the possibility of determining K, say, if η, η_n, H_0, f, f_n are known.

§ 16. Motion in an Elliptical Conduit

Let us consider [14] the same problem which was solved in § 13 (plane conduit) and § 14 (cylindrical conduit) for the case of an elliptical conduit. The steady motion of media with constant true and reduced densities is parallel to the axis of the elliptical cylinder whose cross section is given by the equation

$$\frac{y^2}{b^2} + \frac{z^2}{c^2} - 1 = 0, \tag{16.1}$$

where b and c are the major and minor axes of the ellipse.

Let us write the equations of motion (6.1), which, under the assumptions made above, become

$$f_1\mu_1\left(\frac{\partial^2 u_1}{\partial y^2} + \frac{\partial^2 u_1}{\partial z^2}\right) + K(u_2 - u_1) = f_1\frac{\partial p}{\partial x}, \tag{16.2}$$

$$f_2\mu_2\left(\frac{\partial^2 u_2}{\partial y^2} + \frac{\partial^2 u_2}{\partial z^2}\right) + K(u_1 - u_2) = f_2\frac{\partial p}{\partial x}. \tag{16.3}$$

Combining (16.2) and (16.3) and taking account of (2.1), we obtain

$$\frac{\partial^2 V}{dy^2} + \frac{\partial^2 V}{\partial z^2} = N; \tag{16.4}$$

where

$$V = f_1\mu_1 u_1 + f_2\mu_2 u_2, \ N = \frac{\partial p}{\partial x} = \text{const.}$$

The solution of (16.4) satisfying the boundary condition V = 0 on the ellipse will be

$$V = M\left(\frac{y^2}{b^2} + \frac{z^2}{c^2} - 1\right). \tag{16.5}$$

By inserting this integral into (16.4) we obtain an expression for M, after which (16.5) will become

$$V = \frac{Nb^2c^2}{2(b^2 + c^2)}\left(\frac{y^2}{b^2} + \frac{z^2}{c^2} - 1\right). \tag{16.6}$$

From (16.6) we find the connection between the velocities u_1 and u_2 with the aid of which we obtain the two equations

$$\frac{\partial^2 u_1}{\partial y^2} + \frac{\partial^2 u_1}{\partial z^2} - m^2 u_1 = D - \frac{1}{2}By^2 - \frac{1}{2}Cz^2, \tag{16.7}$$

$$\frac{\partial^2 u_2}{\partial y^2} + \frac{\partial^2 u_2}{\partial z^2} - m^2 u_2 = E - \frac{1}{2}By^2 - \frac{1}{2}Cz^2, \tag{16.8}$$

where

$$\left.\begin{aligned}
m^2 &= K\left(\frac{1}{f_1\mu_1} + \frac{1}{f_2\mu_2}\right) \\
B &= \frac{KNc^2}{(b^2 + c^2)f_1\mu_1 f_2\mu_2} \\
C &= \frac{KNb^2}{(b^2 + c^2)f_1\mu_1 f_2\mu_2} \\
D &= \frac{Cc^2}{2} + \frac{N}{\mu_1} \\
E &= \frac{Cc^2}{2} + \frac{N}{\mu_2}
\end{aligned}\right\} \tag{16.9}$$

Let us proceed to solve (16.7) for the adhesion boundary condition on the ellipse. We seek the solution as

$$u_1 = u_{11} + u_{12},$$
(16.10)

where u_{11} is a particular solution of the inhomogeneous equation, and u_{12} is the general solution of the homogeneous equation. Seeking the particular solution of the inhomogeneous equation u_{11} as a polynomial, we can see that

$$u_{11} = \frac{B}{2m^2} y^2 + \frac{C}{2m^2} z^2 - \left(\frac{D}{2m^2} - \frac{B}{m^4} - \frac{C}{m^4} \right).$$
(16.11)

We then have for u_{12}

$$\frac{\partial^2 u_{12}}{\partial y^2} + \frac{\partial^2 u_{12}}{\partial z^2} - m^2 u_{12} = 0.$$
(16.12)

To solve this equation, let us introduce the elliptic coordinates

$$z = h\,\text{sh}\,\xi \sin \eta, \; y = h\,\text{ch}\,\xi \cos \eta;$$

where

$$h^2 = \sqrt{b^2 - c^2}.$$

In this case

$$\frac{\partial^2 u_{12}}{\partial \xi^2} = \frac{\partial^2 u_{12}}{\partial y^2} h^2 \,\text{sh}^2\,\xi \cos^2 \eta + 2 \frac{\partial^2 u_{12}}{\partial y \partial z} h^2 \text{sh}\xi\text{ch}\xi\sin\eta\cos\eta +$$
$$+ \frac{\partial^2 u_{12}}{\partial z^2} h^2 \text{ch}^2\xi \sin^2\eta + \frac{\partial u_{12}}{\partial y} h\text{ch}\xi\cos\eta + \frac{\partial u_{12}}{\partial z} h\text{sh}\xi\sin\eta$$

$$\frac{\partial^2 u_{12}}{\partial y^2} = \frac{\partial^2 u_{12}}{\partial y^2} h^2 \text{ch}^2\xi \sin^2\eta - 2 \frac{\partial^2 u_{12}}{\partial y \partial z} h^2 \text{sh}\xi\text{ch}\xi\sin\eta\cos\eta +$$
$$+ \frac{\partial^2 u_{12}}{\partial z^2} h^2 \text{sh}^2\xi\sin^2\eta - \frac{\partial u_{12}}{\partial y} h\text{ch}\xi\cos\eta - \frac{\partial u_{12}}{\partial z} h\text{sh}\xi\sin\eta$$
(16.13)

Substituting (16.13) into (16.12) and taking into account that

$$\text{sh}^2\xi\cos^2\eta + \text{ch}^2\xi\sin\eta = \frac{1}{2}(\text{ch}2\xi - \cos2\eta),$$

we have the equation for u_{12} in elliptic coordinates:

$$\frac{\partial^2 u_{12}}{\partial \xi^2} + \frac{\partial^2 u_{12}}{\partial \eta^2} - \frac{h^2 m^2}{2}(\text{ch}2\xi - \cos2\eta) u_{12} = 0.$$
(16.14)

We seek the solution of this equation in the form

$$u_{12} = \varphi(\xi)\,\psi(\eta).$$
(16.15)

Substituting it into (16.14), we obtain

$$\frac{1}{\varphi}\frac{d^2\varphi}{d\xi^2} - \frac{h^2 m^2}{2}\text{ch}\,2\xi = -\frac{1}{\psi}\frac{d^2\psi}{d\eta^2} - \frac{h^2 m^2}{2}\cos^2\eta.$$

Since the left side of this equation is independent of η and the right side is independent of ξ each side should equal a constant a.

We thus obtain two ordinary differential equations

$$\frac{d^2\varphi}{d\xi^2} - (a + 2q\,\mathrm{ch}\,2\xi)\,\varphi = 0, \tag{16.16}$$

$$\frac{d^2\psi}{d\eta^2} + (a + 2q\cos 2\eta)\,\psi = 0, \tag{16.17}$$

where

$$q = \frac{h^2 m^2}{4}.$$

Equations (16.16) and (16.17) are canonical Mathieu equations; the functions satisfying these equations are called Mathieu functions. Hence, the solution of (16.14) consists of the product of two kinds of Mathieu functions, one of which satisfies (16.16) and the other (16.17).

Equation (16.17) has (a) p e r i o d i c   s o l u t i o n s (of the first kind) of the cosine type

$$\mathrm{ce}_{2n}(\eta, -q) = (-1)^n \sum_{r=0}^{\infty} (-1)^r A_{2r}^{(2n)} \cos 2r\eta \tag{16.18}$$

of even order for eigenvalues a_{2n},

$$\mathrm{ce}_{2n+2}(\eta, -q) = (-1)^n \sum_{r=0}^{\infty} (-1)^r B_{2r+1}^{(2n+1)} \cos(2r+1)\eta \tag{16.19}$$

of odd order for eigenvalues b_{2n+1}, and of sine type

$$\mathrm{se}_{(2n+2)}(\eta, -q) = (-1)^n \sum_{r=0}^{\infty} (-1)^r A_{2r+1}^{(2n+1)} \sin(2r+1)\eta \tag{16.20}$$

of odd order for eigenvalues a_{2n+1},

$$\mathrm{se}_{(2n+2)}(\eta, -q) = (-1)^n \sum_{r=0}^{\infty} (-1)^r B_{2r+2}^{(2n+2)} \sin(2r+2)\eta \tag{16.21}$$

of even order for eigenvalues b_{2n+2}; and (b) n o n p e r i o d i c   s o l u t i o n s

$$\mathrm{fe}_{2n+1}(\eta, -q) = -C_{2n}(q)\left[\left(\tfrac{1}{2}\pi, -\eta\right)\mathrm{ce}_{2n}(\eta, -q) + (-1)^n \sum_{r=0}^{\infty} f_{2r+2}^{(2n)} \sin(2r+2)\eta\right],$$

$$\mathrm{fe}_{2n+1}(\eta, -q) = S_{2n+1}(q)\left[\left(\tfrac{1}{2}\pi - \eta\right)\mathrm{ce}_{2n}(\eta, -q) + (-1)^n \sum_{r=0}^{\infty} (-1)^r f_{2r+1}^{(2n+1)} \cos(2r+1)\eta\right],$$

$$\mathrm{ge}_{2n+1}(\eta, -q) = C_{2n+1}(q)\left[\left(\tfrac{1}{2}\pi - \eta\right)\mathrm{se}_{2n+1}(\eta, -q) + (-1)^n \sum_{r=0}^{\infty} (-1)^r f_{2r+1}^{(2n+1)} \cos(2r+1)\eta\right],$$

$$\mathrm{ge}_{2n+2}(\eta, -q) = S_{2n+2}(q)\left[\left(\frac{1}{2}\pi - \eta\right)\mathrm{se}_{2n+2}(\eta, -q) + (-1)^n \sum_{r=0}^{\infty}(-1)^r g_{2r}^{(2n)}\cos 2r\eta\right],$$

where a_{2n}, b_{2n+1}, a_{2n+1}, b_{2n+2} are eigenvalues of the equations; A_m, B_m, f_m, g_m are constants depending on a, q and determined by means of recursion relations.

Equation (16.16) has solutions of the first kind:

$$\mathrm{Ce}_{2n}(\xi, -q) = (-1)^n \sum_{r=0}^{\infty}(-1)^r A_{2r}^{2n}\mathrm{ch}2r\xi \qquad (16.22)$$

for a_{2n};

$$\mathrm{Ce}_{2n+1}(\xi, -q) = (-1)^n \sum_{r=0}^{\infty}(-1)^r B_{2r+1}^{2n+1}\mathrm{ch}(2r+1)\xi \qquad (16.23)$$

for b_{2n+1};

$$\mathrm{Se}_{2n+1}(\xi, -q) = (-1)^n \sum_{r=0}^{\infty}(-1)^r A_{2r+1}^{2n+1}\mathrm{sh}(2r+1)\xi \qquad (16.24)$$

for a_{2n+1};

$$\mathrm{Se}_{2n+1}(\xi, -q) = (-1)^n \sum_{r=0}^{\infty}(-1)^r B_{2r+2}^{2n+2}\mathrm{sh}(2r+2)\xi \qquad (16.25)$$

for b_{2n+2}.

The solution of (16.14) consists of the product of any two functions, each of which is a solution of (16.16) or (16.17) for the same values of a and q.

The flow velocity in each cross section is distributed symmetrically with respect to the major and minor axes of the ellipse. Hence, the products containing the nonperiodic solutions drop out.

The flow expressed by the functions $\mathrm{Ce}_{2n}(\xi, -q)\,\mathrm{ce}_{2n}(\eta, -q)$ will be symmetric relative to both axes. Therefore, this is the single allowable kind of product of Mathieu functions which will satisfy (16.14). Hence, the solution of this equation is represented as

$$u_{12} = \sum_{n=0}^{\infty} C_{1n}\,\mathrm{Ce}_{2n}(\xi, -q)\,\mathrm{ce}_{2n}(\eta, -q). \qquad (16.26)$$

Then

$$u_1 = \frac{Bh^2\,\mathrm{ch}^2\xi\cos^2\eta}{2m^2} + \frac{C\cdot h^2\,\mathrm{sh}^2\xi\sin^2\eta}{2m^2} - \left(\frac{D}{m^2} - \frac{B}{m^4} - \frac{C}{m^4}\right) + \sum_{n=0}^{\infty} C_{1n}\,\mathrm{Ce}_{2n}(\xi, -q)\,\mathrm{ce}_{2n}(\eta, -q). \qquad (16.27)$$

Analogously

$$u_2 = \frac{Bh^2\,\mathrm{ch}^2\xi\cos^2\eta}{2m^2} + \frac{C\cdot h^2\,\mathrm{sh}^2\sin^2\eta}{2m^2} - \left(\frac{E}{m^2} - \frac{B}{m^4} - \frac{C}{m^4}\right) + \sum_{n=0}^{\infty} C_{2n}\,\mathrm{Ce}_{2n}(\xi, -q)\,\mathrm{ce}_{2n}(\eta, -q), \qquad (16.28)$$

where C_{1n}, C_{2n} are constants of integration. Using the boundary conditions, we have

$$\left.\begin{array}{l} \dfrac{N}{m^2\mu_1} - \dfrac{B}{m^4} - \dfrac{C}{m^4} - \displaystyle\sum_{n=0}^{\infty} C_{1n}\,Ce_{2n}\left(\xi_0,\,-q\right)ce_{2n}\left(\eta,\,-q\right) = 0 \\[4mm] \dfrac{N}{m^2\mu_2} - \dfrac{B}{m^4} - \dfrac{C}{m^4} - \displaystyle\sum_{n=0}^{\infty} C_{2n}\,Ce_{2n}\left(\xi_0,\,-q\right)ce_{2n}\left(\eta,\,-q\right) = 0 \end{array}\right\} \tag{16.29}$$

Utilizing the orthogonality property of the Mathieu functions, let us multiply both sides of (16.29) by $ce_{2p}(\eta,\,-q)$ and let us integrate between 0 and 2π with respect to η. Then all the integrals of the products $ce_{2n}(\eta,\,-q)ce_{2p}(\eta,\,-q)$ vanish except for p = n.

Thus

$$\left.\begin{array}{l} \left(\dfrac{N}{m^2\mu_1} - \dfrac{B}{m^4} - \dfrac{C}{m^4}\right)\displaystyle\int_0^{2\pi} ce_{2n}(\eta,-q)\,d\eta = C_{1n}\,Ce_{2n}\left(\xi_0,-q\right)\displaystyle\int_0^{2\pi} ce^2_{2n}(\eta,-q)\,d\eta \\[4mm] \left(\dfrac{N}{m^2\mu_2} - \dfrac{B}{m^4} - \dfrac{C}{m^4}\right)\displaystyle\int_0^{2\pi} ce_{2n}(\eta,-q)\,d\eta = C_{2n}\,Ce_{2n}\left(\xi_0,-q\right)\displaystyle\int_0^{2\pi} ce^2_{2n}(\eta,-q)\,d\eta \end{array}\right\} \tag{16.30}$$

Evaluating the integrals, we find

$$\int_0^{2\pi} ce_{2n}(\eta,\,-q)\,dy = (-1)^n\,2\pi A_0^{(2\pi)},$$

$$\int_0^{2\pi} ce^2_{2n}(\eta,\,-q)\,dy = \pi,$$

where A_0 are Mathieu function coefficients.

We then have from (16.30)

$$C_{1n} = (-1)^n\,\frac{2A_0^{(2n)}}{Ce_{2n}\left(\xi_0,-q\right)}\left(\frac{N}{m^2\mu_1} - \frac{B}{m^4} - \frac{C}{m^4}\right),$$

$$C_{2n} = (-1)^n\,\frac{2A_0^{(2n)}}{Ce_{2n}\left(\xi_0,-q\right)}\left(\frac{F}{m^2\mu_2} - \frac{B}{m^4} - \frac{C}{m^4}\right).$$

Substituting the expressions obtained for the constants of integration into (16.27) and (16.28), we have the final formulas for the phase velocity distribution across an elliptical conduit section:

$$u_1 = \frac{1}{m^2}\left\{\frac{B\,ch^2\xi\cos^2\eta + C\,sh^2\xi\sin^2\eta}{2}\,h^2 - \left(D - \frac{B+C}{m^2}\right) + 2\left(\frac{N}{\mu_1} - \right.\right.$$

$$\left.\left. - \frac{B+C}{m^2}\right)\sum_{n=0}^{\infty}(-1)^n A_0^{2n}\frac{Ce_{2n}(\xi,-q)}{Ce_{2n}(\xi_0,-q)}ce_{2n}(\eta,-q)\right\}, \tag{16.31}$$

$$u_2 = \frac{1}{m^2}\left\{\frac{B\,ch^2\xi\cos^2\eta + C\,sh^2\xi\sin^2\eta}{2}\,h^2 - \right.$$

$$\left. - \left(E - \frac{B+C}{m^2}\right) + 2\left(\frac{N}{\mu_2} - \frac{B+C}{m^2}\right)\sum_{n=0}^{\infty}(-1)^n A_0^{(2n)}\frac{Ce_{2n}(\xi,-q)}{Ce_{2n}(\xi_0,-q)}ce_{2n}(\eta,-q)\right\}. \tag{16.32}$$

Let us consider the transition from an elliptical to a circular conduit. According to the formulas for degeneration of the ordinary modified Mathieu functions [15], we have

$$\mathrm{Ce}_{2n}\,(\xi,\,-q) \to P'_{2n}\,I_{2n}\,(mr),$$

$$\mathrm{Ce}_{2n}\,(\xi_0,\,-q) \to P'_{2n}\,I_{2n}\,(mR);$$

where

$$P'_{2n} = \frac{(-1)^n}{A_0^{(2n)}}\,\mathrm{Ce}_{2n}\,(0,\,-q)\,\mathrm{ce}_{2n}\left(\frac{\pi}{2},\,-q\right);$$

and R is the radius of the circular cylindrical conduit.

The relationship $A_m^{(m)} \to 1$, $B_m^{(m)} \to 1$ is valid for the Mathieu function coefficients of any order as $q \to 0$, while all the remaining coefficients $A_p^{(m)}$ and $B_p^{(m)}$ tend to zero. It hence follows that

$$\mathrm{ce}_{2n}\,(\eta,\,-q) \to \cos 2n\eta,$$

$$\mathrm{ce}_0\,(\eta,\,-q) \to A_0^{(0)} = 2^{-\frac{1}{2}},$$

$$A_0^{(2n)} \to 0.$$

Moreover, we have

$$c \to b \to R,$$
$$y = h\,\mathrm{ch}\,\xi\cos\eta \to r\cos\eta,$$
$$z = h\,\mathrm{sh}\,\xi\sin\eta \to r\sin\eta.$$

Therefore, it is easy to see that (16.27), (16.28) take the form of the expressions (14.20) and (14.21).

For $\mu_1 = \mu_2 = \mu$ we obtain the velocity distribution of a single-phase fluid over the cross section [16]:

$$u = \frac{b^2 c^2}{2\,(b^2 + c^2)\,\mu}\,\frac{\partial p}{\partial x}\left(\frac{y^2}{b^2} + \frac{z^2}{c^2} - 1\right).$$

Let us define the discharge of the media by means of the formulas

$$\left.\begin{array}{l} Q_1 = f_1 \iint\limits_s u_1\,(\xi,\,\eta)\,ds \\[2mm] Q_2 = f_2 \iint u_2\,(\xi,\,\eta)\,ds \end{array}\right\}. \tag{16.33}$$

Since the differentials of the arcs are

$$ds_1 = \sqrt{\left(\frac{\partial y}{\partial \xi}\right)^2 + \left(\frac{\partial z}{\partial \xi}\right)^2}\,d\xi,$$

$$ds_2 = \sqrt{\left(\frac{\partial y}{\partial \eta}\right)^2 + \left(\frac{\partial z}{\partial \eta}\right)^2}\,d\eta,$$

then

$$ds_1 = l_1 d\xi, \quad ds_2 = l_2 d\eta,$$

where

$$l_1 = l_2 = \frac{h}{\sqrt{2}} \sqrt{\operatorname{ch} 2\xi - \cos 2\eta}.$$

Then

$$Q_1 = \frac{f_1 h^2}{2} \int\limits_{-\xi_0}^{\xi_0} \int\limits_{0}^{2\pi} u_1(\xi, \eta)(\operatorname{ch} 2\xi - \cos 2\eta)\, d\xi d\eta,$$

$$Q_2 = \frac{f_2 h^2}{2} \int\limits_{-\xi_0}^{\xi_0} \int\limits_{0}^{2\pi} u_2(\xi, \eta)(\operatorname{ch} 2\xi - \cos 2\eta)\, d\xi d\eta.$$

Integrating these expressions, then substituting the values of m, B, C, D, E, and taking into account that

$$b = h \operatorname{ch} \xi_0 \quad c = h \operatorname{sh} \xi_0,$$

we obtain

$$Q_1 = -\frac{f_1 \pi bc}{m^2} \frac{\partial p}{\partial x} \left\{ \frac{b^2 c^2}{4(b^2 + c^2) f_1 \mu_1 f_2 \mu_2} + \frac{1}{\mu_1} - \frac{1}{f_1 \mu_1 + f_2 \mu_2} - \left(\frac{1}{\mu_1} - \frac{1}{f_1 \mu_1 + f_2 \mu_2} \right) \sum_{n=0}^{\infty} (-1)^n A_0^{(2n)} \left[2 \left(A_0^{(2n)} \right)^2 bc + \right. \right.$$
$$\left. \left. + \sum_{r=1}^{\infty} (-1)^{r+1} \frac{h^2}{\eta} \left(A_0^{(2n)} A_{2r+2}^{(2n)} - A_2^{(2n)} A_{2r}^{(2n)} + A_0^{(2n)} A_{2r+2}^{(2n)} \right) \operatorname{sh} 2r \left(\operatorname{arcch} \frac{b}{h} \right) \right] \frac{1}{\operatorname{Ce}_{2n}\left(\operatorname{arcch}\frac{b}{h}, -q \right)} \right\}, \quad (16.34)$$

$$Q_2 = -\frac{f_2 \pi bc}{m^2} \frac{\partial p}{\partial x} \left\{ \frac{b^2 c^2}{4(b^2 + c^2) f_1 \mu_1 f_2 \mu_2} + \frac{1}{\mu_2} - \frac{1}{f_1 \mu_1 + f_2 \mu_2} - \left(\frac{1}{\mu_2} - \frac{1}{f_1 \mu_1 + f_2 \mu_2} \right) \sum_{n=0}^{\infty} (-1)^n A_0^{(2n)} \left[2 \left(A_0^{(2n)} \right)^2 bc + \right. \right.$$
$$\left. \left. + \sum_{r=1}^{\infty} (-1)^{r+1} \frac{h^2}{\eta} \left(A_0^{(2n)} A_{2r+2}^{(2n)} - A_2^{(2n)} A_{2r}^{(2n)} + A_0^{(2n)} A_{2r+2}^{(2n)} \right) \operatorname{sh} 2r \left(\operatorname{arcch} \frac{b}{h} \right) \right] \frac{1}{\operatorname{Ce}_{2n}\left(\operatorname{arcch}\frac{b}{h}, -q \right)} \right\}. \quad (16.35)$$

If the ellipse goes over into a circle, then the preceding formulas of the discharges take the form of (14.22) and (14.23) derived for the circular conduit.

It is known that in the case of single-phase fluid motion in conduits, the optimum cross section in terms of energy capacity is the circle. However, the circular section will not be optimum for the motion of two-phase media taking account of gravity (for fine solid particles).

A. I. Kuprin and Chen Da-jun [17] studied the motion of a two-phase flow in an elliptical conduit experimentally and proved the advantage of this section over the circular, but this has not yet been worked out theoretically. To resolve this question it is necessary to solve the problem of the motion of a two-phase flow taking account of gravity and the variability of the porosity along the vertical in a circular conduit, as well as to compare the energy capacity of two-phase flow motion in the circular and the elliptical conduits of equal cross sections for solid discharges.

The above-mentioned problem may be posed more broadly if it is assumed that the two-phase flow moves in a conduit with cross section $F(x, y)$ (the $0z$ axis coincides with the conduit axis perpendicular to the x, y plane). It is then necessary to find that shape of the cross section for which the energy capacity would be least for a given solid discharge and cross-sectional area.

§ 17. Motion in a Circular Coaxial Annular Conduit

The equations of motion in a cylindrical coordinate system for rectilinear-parallel steady motion of incompressible viscous two-phase media with constant porosity possessing axial symmetry have the form of (14.2), (14.3), as has been shown in § 14.

Let us consider the motion in a coaxial circular annular conduit [18]. We direct the $0x$ axis along the conduit axis.

Let a and b be the radii of the coaxial conduits; then to determine the velocity distribution of the media over the cross section we have (14.2), (14.3), and the boundary conditions

$$\left. \begin{array}{l} u_1 = u_2 = 0 \quad \text{for} \quad r = a \\ u_1 = u_2 = 0 \quad \text{for} \quad r = b \end{array} \right\}. \tag{17.1}$$

Term-by-term addition of (14.2) and (14.3), taking account of (2.1), yields

$$f_1\mu_1 \left(\frac{d^2u_1}{dr^2} + \frac{1}{r}\frac{du_1}{dr} \right) + f_2\mu_2 \left(\frac{d^2u_2}{dr^2} + \frac{1}{r}\frac{du_2}{dr} \right) = N.$$

After integrating this equation under the boundary conditions (17.1) we have

$$f_1\mu_1 u_1 + f_2\mu_2 u_2 = \frac{N}{4} r^2 - \frac{N(a^2 - b^2)}{4 \ln \frac{a}{b}} \ln r + \frac{N(a^2 \ln b - b^2 \ln a)}{4 \ln \frac{a}{b}}. \tag{17.2}$$

Equation (17.2) is a dependence between the velocities of the media. First solving for u_2, then for u_1, and substituting the values found into (14.2) and (14.3), we obtain

$$\frac{d^2u_1}{dr^2} + \frac{1}{r}\frac{du_1}{dr} - m^2 u_1 = Ar^2 + B + C \ln r, \tag{17.3}$$

$$\frac{d^2u_2}{dr^2} + \frac{1}{r}\frac{du_2}{dr} - m^2 u_2 = Ar^2 + D + C \ln r, \tag{17.4}$$

where

$$\left. \begin{array}{l} A = -\dfrac{KN}{4f_1\mu_1 f_2\mu_2} \\[4mm] B = N\left[\dfrac{1}{\mu_1} - \dfrac{K(a^2 \ln b - b^2 \ln a)}{4f_1\mu_1 f_2\mu_2 \ln \dfrac{a}{b}} \right] \\[6mm] D = N\left[\dfrac{1}{\mu_2} - \dfrac{K(a^2 \ln b - b^2 \ln a)}{4f_1\mu_1 f_2\mu_2 \ln \dfrac{a}{b}} \right] \\[6mm] C = \dfrac{KN(a^2 - b^2)}{4 f_1\mu_1 f_2\mu_2 \ln \dfrac{a}{b}} \\[5mm] m^2 = K\left(\dfrac{1}{f_1\mu_1} + \dfrac{1}{f_2\mu_2} \right) \end{array} \right\} \tag{17.5}$$

The differential equations (17.3) and (17.4) differ only in the given constants B and D; hence, it is sufficient to solve just (17.3). Let us substitute $\xi = mr$ whereupon (17.3) becomes

$$\frac{d^2 u_1}{d\xi^2} + \frac{1}{\xi}\frac{du_1}{d\xi} - u_1 = \frac{A}{m^4}\xi^2 + \frac{B}{m^2} + \frac{C}{m^2}\ln\frac{\xi}{m}. \tag{17.6}$$

We solve this last equation by the Lagrange method.

As is known, the homogeneous equation

$$\frac{d^2 u_1}{d\xi^2} + \frac{1}{\xi}\frac{du_1}{d\xi} - u_1 = 0,$$

has the solution

$$u_1^* = c_1 I_0(\xi) + c_2 K_0(\xi). \tag{17.7}$$

Considering c_1 and c_2 variable, according to the Lagrange method we may write

$$c_1' I_0(\xi) + c_2' K_0(\xi) = 0,$$

$$c_1' I_0'(\xi) + c_2' K_0'(\xi) = \frac{A}{m^4}\xi^2 + \frac{B}{m^2} + \frac{C}{m^2}\ln\frac{\xi}{m}.$$

In this case

$$c_1 = \frac{1}{m^2}\int \frac{\left(\frac{A}{m^2}\xi^2 + B + C\ln\frac{\xi}{m}\right)K_0(\xi)\,d\xi}{I_0'(\xi)K_0(\xi) - K_0'(\xi)I_0(\xi)},$$

$$c_2 = -\frac{1}{m^2}\int \frac{\left(\frac{A}{m^2}\xi^2 + B + C\ln\frac{\xi}{m}\right)I_0(\xi)\,d\xi}{I_0'(\xi)K_0(\xi) - K_0'(\xi)I_0(\xi)},$$

or

$$\left.\begin{aligned}
c_1 &= \frac{1}{m^2}\int\left(\frac{A}{m^2}\xi^3 + B\xi + C\xi\ln\frac{\xi}{m}\right)K_0(\xi)\,d\xi \\
c_2 &= -\frac{1}{m^2}\int\left(\frac{A}{m^2}\xi^3 + B\xi + C\xi\ln\frac{\xi}{m}\right)I_0(\xi)\,d\xi
\end{aligned}\right\} \tag{17.8}$$

Evaluating the integrals

$$\int \xi\ln\frac{\xi}{m}K_0(\xi)\,d\xi = -\xi K_1(\xi)\ln\frac{\xi}{m} - K_0(\xi),$$

$$\int \xi\ln\frac{\xi}{m}I_0(\xi)\,d\xi = \xi I_1(\xi)\ln\frac{\xi}{m} - I_0(\xi),$$

we obtain

$$c_1 = -\frac{A}{m^4}\xi^3 K_1(\xi) - \frac{2A}{m^4}\xi^2 K_0(\xi) - \left(\frac{4A}{m^4} + \frac{B}{m^2}\right)\xi K_1(\xi) - \frac{C}{m^2}\xi K_1(\xi)\ln\frac{\xi}{m} - \frac{C}{m^2}K_0(\xi) + C_{01},$$

$$c_2 = -\frac{A}{m^4}\xi^3 I_1(\xi) + \frac{2A}{m^4}\xi^2 I_0(\xi) - \left(\frac{4A_4}{m^4} + \frac{B_4}{m^2}\right)\xi I_1(\xi) - \frac{C}{m^2}\xi I_1(\xi)\ln\frac{\xi}{m} + \frac{C}{m^2}I_0(\xi) + c_{02}.$$

Substituting the values of c_1 and c_2 into (17.7) and taking account of (14.13) we obtain

$$u_1 = c_{11}I_0(\xi) + c_{12}K_0(\xi) - \frac{A_4}{m^4}\xi^2 - \left(\frac{4A}{m^4} + \frac{B}{m^2}\right) - \frac{C}{m^2}\ln\frac{\xi}{m}\,.$$

Transforming from ξ to the previous variable r, we rewrite the last equation as

$$u_1 = c_{11}I_0(mr) + c_{12}K_0(mr) - \frac{A}{m^2}r^2 - \left(\frac{4A}{m^4} + \frac{B}{m^2}\right) - \frac{C}{m^2}\ln r. \tag{17.9}$$

Replacing B by D in (17.9), we have an expression for the velocity of the second medium:

$$u_2 = c_{13}I_0(mr) + c_{14}K_0(mr) - \frac{A}{m^2}r^2 - \left(\frac{4A}{m^4} - \frac{D}{m^2}\right) - \frac{C}{m^2}\ln r. \tag{17.10}$$

The c_{11}, c_{12}, c_{13}, c_{14} in (17.9) and (17.10) are constants of integration which can be determined by utilizing the boundary condition (17.1):

$$c_{11} = \frac{1}{I_0(ma)K_0(mb) - K_0(ma)I_0(mb)}\left[K_0(mb)\left(\frac{Aa^2}{m^2} + \frac{4A}{m^4} + \frac{B}{m^2} + \frac{C\ln a}{m^2}\right) - \right.$$

$$\left. - K_0(ma)\left(\frac{Ab^2}{m^2} + \frac{4A}{m^4} + \frac{B}{m^2} + \frac{C\ln b}{m^2}\right)\right], \tag{17.11}$$

$$c_{12} = \frac{1}{K_0(ma)I_0(mb) - I_0(ma)K_0(mb)}\left[I_0(mb)\left(\frac{Aa^2}{m^2} + \frac{4A}{m^4} + \frac{B}{m^2} + \frac{C\ln a}{m^2}\right) - \right.$$

$$\left. - I_0(ma)\left(\frac{Ab^2}{m^2} + \frac{4A}{m^4} + \frac{B}{m^2} + \frac{C\ln b}{m^2}\right)\right]. \tag{17.12}$$

The values of c_{13} and c_{14} are obtained after B has been replaced by D in the expressions for c_{11} and c_{12}. Substituting these expressions into (17.9), we obtain

$$u_1 = -\frac{N\left[\frac{1}{f_1\mu_1 + f_2\mu_2} - \frac{1}{\mu_1}\right]}{[I_0(ma)K_0(mb) - K_0(ma)I_0(mb)]m^2}\left\{I_0(mr)\left[K_0(mb) - K_0(ma)\right] - K_0(mr)\left[I_0(mb) - I_0(ma)\right]\right\} -$$

$$- N\left\{-\frac{1}{4(f_1\mu_1 + f_2\mu_2)}r^2 - \frac{1}{(f_1\mu_1 + f_2\mu_2)m^2} + \frac{1}{m^2\mu_1} - \frac{a^2\ln b - b^2\ln a}{4(f_1\mu_1 + f_2\mu_2)\ln\frac{a}{b}} + \frac{(a^2 - b^2)\ln r}{4(f_1\mu_1 + f_2\mu_2)\ln\frac{a}{b}}\right\}, \tag{17.13}$$

$$u_2 = -\frac{N\left[\frac{1}{f_1\mu_1 + f_2\mu_2} - \frac{1}{\mu_2}\right]}{[I_0(ma)K_0(mb) - K_0(ma)I_0(mb)]m^2}\left\{I_0(mr)\left[K_0(mb) - K_0(ma)\right] - K_0(mr)\times\right.$$

$$\times\left[I_0(mb) - I_0(ma)\right]\right\} - N\left\{-\frac{1}{4(f_1\mu_1 + f_2\mu_2)}r^2 - \right.$$

$$\left. - \frac{1}{(f_1\mu_1 + f_2\mu_2)m^2} + \frac{1}{m^2\mu_2} - \frac{a^2\ln b - b^2\ln a}{4(f_1\mu_1 + f_2\mu_2)\ln\frac{a}{b}} - \frac{(a^2 - b^2)\ln r}{4(f_1\mu_1 + f_2\mu_2)\ln\frac{a}{b}}\right\}. \tag{17.14}$$

Expressions (17.13) and (17.14) picture the velocity distribution over the cross section of a circular coaxial annular conduit.

We determine the discharge of the media by means of the formulas

$$Q_1 = f_1\int_b^a 2\pi r u_1 dr,$$

$$Q_2 = f_2 \int_b^a 2\pi r u_2 dr.$$

Substituting the values for u_1 and u_2 from (17.13) and (17.14) into these formulas and integrating between b and a, taking into account that

$$\int r I_0 (mr) \, dr = \frac{r}{m} I_1 (mr),$$

$$\int r K_0 (mr) \, dr = - \frac{r}{m} K_1 (mr),$$

we obtain

$$Q_1 = - \frac{2\pi f_1 N \left(\frac{1}{f_1\mu_1 + f_2\mu_2} - \frac{1}{\mu_1} \right)}{[I_0 (ma) K_0 (mb) - K_0 (ma) I_0 (mb)] \, m^3} \Big\{ a [K_0 (mb) I_1 (ma) + K_1 (ma) I_0 (mb)] +$$

$$+ b [K_0 (ma) I_1 (mb) + K_1 (mb) I_0 (ma)] - \frac{2}{m} \Big\} -$$

$$- \pi f_1 N \left[\frac{a^4 - b^4}{8 (f_1\mu_1 + f_2\mu_2)} - \frac{(a^2 - b^2)^2}{8 (f_1\mu_1 + f_2\mu_2) \ln \frac{a}{b}} - \frac{a^2 - b^2}{(f_1\mu_1 + f_2\mu_2) \, m^2} + \frac{a^2 - b^2}{m^2\mu_1} \right], \qquad (17.15)$$

$$Q_2 = - \frac{2\pi f_2 N \left(\frac{1}{f_1\mu_1 + f_2\mu_2} - \frac{1}{\mu_2} \right)}{[I_0 (ma) K_0 (mb) - K_0 (ma) I_0 (mb)] m^3} \Big\{ a [K_0 (mb) I_1 (ma) + K_1 (ma) I_0 (mb)] +$$

$$+ b [K_0 (ma) I_1 (mb) + K_1 (mb) I_0 (ma)] - \frac{2}{m} \Big\} -$$

$$- \pi f_2 N \left[\frac{a^4 - b^4}{8 (f_1\mu_1 + f_2\mu_2)} - \frac{(a^2 - b^2)^2}{8 (f_1\mu_1 + f_2\mu_2) \ln \frac{a}{b}} - \frac{a^2 - b^2}{(f_1\mu_1 + f_2\mu_2) \, m^2} + \frac{a^2 - b^2}{m^2\mu_2} \right], \qquad (17.16)$$

which are formulas for the discharge of the media moving in a two-phase mixture.

Let us note that for $\mu_1 = \mu_2 = \mu$ we obtain the formulas well known in hydrodynamics for a homogeneous incompressible fluid [10]:

$$u = - \frac{N}{4\mu} \left[\frac{a^2 - b^2}{\ln \frac{a}{b}} \ln r - r^2 - \frac{a^2 \ln b - b^2 \ln a}{\ln \frac{a}{b}} \right],$$

$$Q = Q_1 + Q_2 = - \frac{N\pi}{8\mu} \left[a^4 - b^4 - \frac{(a^2 - b^2)^2}{\ln \frac{a}{b}} \right].$$

§ 18. Motion of Two-Phase Media in the Space between Piping

Let us consider the motion of two-phase media in a conduit within which small pipes of the same radius σ are disposed in an unstaggered manner. Such motion is often encountered in the petroleum industry. In order to avoid solidification of the petroleum, small pipes are used as heaters; a cross section of the arrangement is shown in Fig. 4.

Because petroleum contains impurities, it is interesting [19] to solve this problem from the viewpoint of multiphase media.

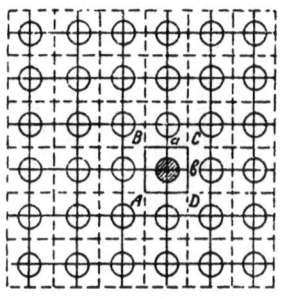
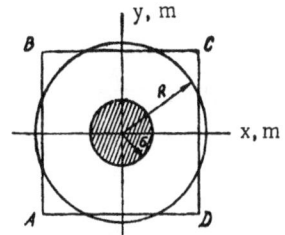

Fig. 4   Fig. 5

Let us study the rectilinear-parallel steady flow of incompressible viscous two-phase media without taking account of temperature changes. Under these conditions the equations of motion in cylindrical coordinates take the form of the system (14.1).

Let the motion be in the space between two cylindrical pipes in a direction parallel to the pipe axes and coincident with the 0x axis.

Because of symmetry, the flows in all the dashed rectangles in Fig. 4 are the same; hence it is sufficient to consider just one such element (Fig. 5). Because of adhesion of the media to the pipe walls the boundary conditions at r = σ will be

$$u_1 = 0, \; u_2 = 0.$$ (18.1)

The velocities of the media at all points of the element's outline should be perpendicular to a normal drawn to the outline, i.e., on ABCD

$$\frac{du_1}{dn} = \frac{du_2}{dn} = 0.$$ (18.2)

To simplify the solution of the problem, let us replace the area of the rectangle ABCD by the area of an equivalent circle of radius

$$R = \sqrt{\frac{ab}{\pi}},$$

where a, b are the sides of the rectangle. Then for r = R (18.2) becomes

$$\frac{du_1}{dr} = \frac{du_2}{dr} = 0.$$ (18.3)

Moreover, if $a = b$,

$$R = \frac{a}{\sqrt{\pi}} = 0.5642a.$$

Because of axial symmetry ($\partial u_1 / \partial \varphi = 0$, $\partial u_2 / \partial \varphi = 0$), equations (14.1) take the form of (14.2) and (14.3).

Therefore, the problem has been reduced to the solution of the system (14.2) and (14.3) under the boundary conditions (18.1) and (18.3). Combining (14.2) and (14.3) and taking account of (2.1) we have

$$f_1 \mu_1 \frac{d}{dr}\left(r \frac{du_1}{dr}\right) + f_2 \mu_2 \frac{d}{dr}\left(r \frac{du_2}{dr}\right) = Nr.$$

Integrating this equation once, utilizing condition (18.3), and then integrating again under conditions (18.1), we obtain

$$f_1\mu_1 u_1 + f_2\mu_2 u_2 = \frac{N}{4}r^2 - \frac{NR^2}{2}\ln r + \frac{NR^2}{4}\left[2\ln\sigma - \left(\frac{\sigma}{R}\right)^2\right],$$

from which we find

$$u_2 = \frac{N}{4f_2\mu_2}r^2 - \frac{NR^2}{2f_2\mu_2}\ln r + \frac{NR^2}{4f_2\mu_2}\left[2\ln\sigma - \left(\frac{\sigma}{R}\right)^2\right] - \frac{f_1\mu_1}{f_2\mu_2}u_1,$$

$$u_1 = \frac{N}{4f_1\mu_1}r^2 - \frac{NR^2}{2f_1\mu_1}\ln r + \frac{NR^2}{4f_1\mu_1}\left[2\ln\sigma - \left(\frac{\sigma}{R}\right)^2\right] - \frac{f_2\mu_2}{f_1\mu_1}u_2.$$

Inserting these expressions into (14.2) and (14.3), we find

$$\frac{d^2u_1}{dr^2} + \frac{1}{r}\frac{du_1}{dr} - m^2u_1 = Ar^2 + B\ln r + C, \tag{18.4}$$

$$\frac{d^2u_2}{dr^2} + \frac{1}{r}\frac{du_2}{dr} - m^2u_2 = Ar^2 + B\ln r + D, \tag{18.5}$$

where

$$\left.\begin{aligned}
m^2 &= K\left(\frac{1}{f_1\mu_1} + \frac{1}{f_2\mu_2}\right) \\[4pt]
A &= -\frac{KN}{4f_1\mu_1 f_2\mu_2} \\[4pt]
B &= \frac{KNR^2}{2f_1\mu_1 f_2\mu_2} \\[4pt]
C &= \frac{N}{\mu_1} - \frac{KNR^2}{4f_1\mu_1 f_2\mu_2}\left[2\ln\sigma - \left(\frac{\sigma}{R}\right)^2\right] \\[4pt]
D &= \frac{N}{\mu_2} - \frac{KNR^2}{4f_1\mu_1 f_2\mu_2}\left[2\ln\sigma - \left(\frac{\sigma}{R}\right)^2\right]
\end{aligned}\right\} \tag{18.6}$$

The expressions (18.4) and (18.5) are the Bessel equation with a right side. Let us solve (18.4) by the Lagrange method under the boundary conditions (18.1) and (18.3). It is known that the general solution of the homogeneous equation

$$\frac{d^2u_1}{dr^2} + \frac{1}{r}\frac{du_1}{dr} - m^2u_1 = 0$$

is

$$u_1 = c_1 I_0(mr) + c_2 K_0(mr). \tag{18.7}$$

Let us select the desired c_1 and c_2 so that the equalities

$$c_1' I_0(mr) + c_2' K_0(mr) = 0,$$

$$c_1' I_0(mr) + c_2' K_0(mr) = \frac{Ar^2 + B\ln r + C}{m}$$

are satisfied. From the preceding system we obtain integrals to find the c_1 and c_2 evaluated by the method examined in § 17. We then have

$$u_1 = c_{11} I_0(mr) + c_{12} K_0(mr) - \frac{A}{m^2} r^2 - \frac{4A}{m^4} - \frac{c}{m^2} - \frac{B}{m^2} \ln r. \tag{18.8}$$

Similarly, we find

$$u_2 = c_{21} I_0(mr) + c_{22} K_0(mr) - \frac{A}{m^2} r^2 - \frac{4A}{m^4} - \frac{D}{m^2} - \frac{B}{m^2} \ln r, \tag{18.9}$$

where the c_{11}, c_{12}, c_{21}, c_{22} are constants of integration which can be determined from the boundary conditions (18.1) and (18.3):

$$c_{11} = \frac{1}{I_0(m\sigma) K_1(mR) + K_0(m\sigma) I_1(mR)} \left[\frac{K_1(mR)}{m^2} \left(A\sigma^2 + \frac{4A}{m^2} + C + B \ln \sigma \right) + \frac{K_0(m\sigma)}{m} \left(2AR + \frac{B}{R} \right) \right], \tag{18.10}$$

$$c_{12} = \frac{1}{K_0(m\sigma) I_1(mR) + I_0(m\sigma) K_1(mR)} \left[\frac{I_1(mR)}{m^2} \left(A\sigma^2 + \frac{4A}{m^2} + C + B \ln \sigma \right) - \frac{I_0(m\sigma)}{m} \left(2AR + \frac{B}{R} \right) \right]. \tag{18.11}$$

The values of c_{21} and c_{22} are obtained by replacing C by D in the equations for c_{11} and c_{12}.

The discharge for each medium is determined by means of the formulas

$$Q_1 = f_1 \int_\sigma^R 2\pi r u_1 dr,$$

$$Q_2 = f_2 \int_\sigma^R 2\pi r u_2 dr.$$

Inserting the values for u_1 and u_2 in the preceding formulas and integrating, we have

$$Q_1 = \frac{2\pi f_1}{m} \left\{ \frac{RI_1(mR) - \sigma I_1(m\sigma)}{I_0(m\sigma) K_1(mR) + K_0(m\sigma) I_1(mR)} \left[\frac{K_1(mR)}{m^2} \left(A\sigma^2 + \frac{4A}{m^2} + C + B \ln \sigma \right) + \frac{K_0(m\sigma)}{m} \left(2AR + \frac{R}{2} \right) \right] - \right.$$

$$- \frac{RK_1(mR) - \sigma K_1(m\sigma)}{K_0(m\sigma) I_1(mR) + I_0(m\sigma) K_1(mR)} \left[\frac{I_1(mR)}{m^2} \left(A\sigma^2 + \frac{4A}{m^2} + C + B \ln \sigma \right) - \frac{I_0(mR)}{m} \left(2AR + \frac{B}{R} \right) \right] -$$

$$\left. - \frac{A}{4m} (R^4 - \sigma^4) + \frac{R^2 - \sigma^2}{2m} \left(\frac{B}{2} - \frac{4A}{m^2} - C \right) - \frac{B}{2m} (R^2 \ln R - \sigma^2 \ln \sigma) \right\}. \tag{18.12}$$

If D replaces C in this formula, we obtain a formula for the discharge of the second phase Q_2.

Let us find the average velocity of the flow for the first medium, say, from the equality

$$Q_1 = f_1 F u_{1m}.$$

where $F = \pi(R^2 - \sigma^2)$ is the cross-sectional area of the element considered. Hence, according to (18.12), we have

$$u_{1m} = \frac{2N}{\mu_1 m^2 (R^2 - \sigma^2)} \left\{ \frac{RI_1(mR) - \sigma I_1(m\sigma)}{I_0(m\sigma) K_1(mR) + K_0(m\sigma) I_1(mR)} \left[\frac{K_1(mR)}{m} \left(1 - \frac{K}{m^2 f_1 f_2 \mu_2} \right) \right] - \right.$$

$$- \frac{RK_1(mR) - \sigma K_1(mR)}{K_0(m\sigma) I_1(mR) + I_0(m\sigma) K_1(mR)} \left[\frac{I_1(mR)}{m} \left(1 - \frac{K}{m^2 f_1 f_2 \mu_2} \right) \right] + \frac{K}{4 f_1 f_2 \mu_2} \left[\frac{R^4 - \sigma^4}{4} - \right.$$

$$\left. - R^2 (R^2 \ln R - R^2 \ln \sigma) \right] + \frac{K(R^2 - \sigma^2)}{2 f_1 f_2 \mu_2} \left[\frac{R^2 - \sigma^2}{4} + \frac{2}{m^2} - \frac{f_1 f_2 \mu_2}{K} \right] \right\}. \tag{18.13}$$

We find the dependence of the Reynolds number for the first phase from the other parameters. We define this number in the same manner as for single-phase motion:

$$\mathrm{Re} = \frac{4u_{1m}\delta}{\nu_1} \, , \tag{18.14}$$

where δ is the hydraulic radius, and ν_1 is the coefficient of kinematic viscosity of the first medium.

The wetted perimeter is just the perimeter of the pipe; hence, the ratio of area of the active element to the length of the wetted perimeter yields

$$\delta = \frac{f_1(R^2 - \sigma^2)}{2\sigma} \, ,$$

whence (18.14) becomes

$$\mathrm{Re}' = \frac{2u_{1m}f_1(R^2 - \sigma^2)}{\sigma \nu_1} \, . \tag{18.15}$$

If $f_2 = 0$, from (18.15) an expression for the Reynolds number of a single-phase fluid is obtained (i.e., for the first phase in the absence of the second phase):

$$\mathrm{Re} = \frac{2u_m(R^2 - \sigma^2)}{\sigma \nu} \, . \tag{18.16}$$

Since $\nu_1 = \nu$, on the basis of (18.15) and (18.16) we obtain

$$\frac{\mathrm{Re}'}{\mathrm{Re}} = \alpha, \tag{18.17}$$

where

$$\alpha = f_1 \frac{u_{1m}}{u_m}. \tag{18.18}$$

Knowledge of these last two dependences permits the determination of the influence of impurities on the Reynolds number of a fluid, on which many factors depend (the drag of the conduit, etc.).

Let us note that under the conditions $\mu_1 = \mu_2 = \mu$ the formulas for the velocities and the total discharge take the well-known single-phase flow form [16]

$$u = \frac{N}{4\mu}(\sigma^2 - r^2) + \frac{NR^2}{2\mu}\ln\frac{r}{\sigma} \, ,$$

$$Q = -\frac{N\pi(R^2 - \sigma^2)^2}{8\mu} + \frac{N\pi R^2}{2\mu}\left[R^2\ln\frac{R}{\sigma} - \frac{R^2 - \sigma^2}{2}\right].$$

§ 19. Motion between Fixed Arcs of Concentric Circles

In the general case of circular motion all the velocity components of the media $v_{1\varphi}$, $v_{2\varphi}$, v_{1r}, v_{2r}, w_1, w_2 must be considered different from zero. Moreover, the porosities will be variable. Under such conditions the stratification of media entering a conduit in the form of a mixture may be studied. However, in certain cases some simplifications can be made for a large radius of curvature or for a slight difference in the densities of the media, and an exact solution of the problem can be obtained.

Thus, let the velocity components of the media along the radius and along the axis be zero;

$$v_{1r} = v_{2r} = 0,$$
$$w_1 = w_2 = 0$$

and let the porosities be constant. Then the continuity equations yield

$$\frac{\partial v_{1\varphi}}{\partial \varphi} = 0, \qquad \frac{\partial v_{2\varphi}}{\partial \varphi} = 0. \tag{19.1}$$

The system of differential equations (8.3) is hence rewritten as [20]

$$\left.\begin{aligned}
\rho_1 \frac{v_{1\varphi}^2}{2} &= f_1 \frac{\partial p}{\partial r} \\[2mm]
\rho_2 \frac{v_{2\varphi}^2}{2} &= f_2 \frac{\partial p}{\partial r} \\[2mm]
\frac{\partial p}{\partial z} &= 0
\end{aligned}\right\} \tag{19.2}$$

$$\left.\begin{aligned}
f_1 \mu_1 \left(\frac{\partial^2 v_{1\varphi}}{\partial r^2} + \frac{\partial^2 v_{1\varphi}}{\partial z^2} + \frac{1}{r} \frac{\partial v_{1\varphi}}{\partial r} - \frac{v_{1\varphi}}{r^2} \right) + K\left(v_{2\varphi} - v_{1\varphi} \right) = f_1 \frac{1}{r} \frac{\partial p}{\partial \varphi} \\[2mm]
f_2 \mu_2 \left(\frac{\partial^2 v_{2\varphi}}{\partial r^2} + \frac{\partial^2 v_{2\varphi}}{\partial z^2} + \frac{1}{r} \frac{\partial v_{2\varphi}}{\partial r} - \frac{v_{2\varphi}}{r^2} \right) + K\left(v_{1\varphi} - v_{2\varphi} \right) = f_2 \frac{1}{r} \frac{\partial p}{\partial \varphi}
\end{aligned}\right\} \tag{19.3}$$

Differentiating the first two equations of the system (19.2) with respect to z and taking account of the third equation of this system, we have

$$\frac{\partial v_{1\varphi}}{\partial z} = 0, \qquad \frac{\partial v_{2\varphi}}{\partial z} = 0.$$

This means that in the case under consideration the circular motion of viscous two-phase media should be considered plane-parallel. Then (19.3) may be rewritten thus:

$$f_1 \mu_1 \left(\frac{d^2 v_{1\varphi}}{dr^2} + \frac{1}{r} \frac{\partial v_{1\varphi}}{\partial r} - \frac{v_{1\varphi}}{r^2} \right) + K\left(v_{2\varphi} - v_{1\varphi} \right) = \frac{f_1 N}{r}, \tag{19.4}$$

$$f_2 \mu_2 \left(\frac{d^2 v_{2\varphi}}{dr^2} + \frac{1}{r} \frac{dv_{2\varphi}}{dr} - \frac{v_{2\varphi}}{r^2} \right) + K\left(v_{1\varphi} - v_{2\varphi} \right) = \frac{f_2 N}{r}, \tag{19.5}$$

where

$$N = \frac{\partial p}{\partial \varphi} = \text{const.}$$

Let the motion be between arcs of concentric circles of radii a and b. In order to derive the formulas for the velocity distribution over the cross section we have equations (19.4), (19.5), and the boundary conditions

$$\left.\begin{aligned}
v_{1\varphi} &= 0, \ v_{2\varphi} = 0 \quad \text{for} \quad r = b \\[2mm]
v_{1\varphi} &= 0, \ v_{2\varphi} = 0 \quad \text{for} \quad r = a
\end{aligned}\right\} \tag{19.6}$$

Addition of (19.4) and (19.5) term by term and taking account of (2.1) yields

$$\frac{d}{dr}\left[\frac{1}{r} \frac{d}{dr} (Mr) \right] = \frac{N}{r}, \tag{19.7}$$

where

$$M = f_1 \, \mu_1 \, v_{1\varphi} + f_2 \, \mu_2 \, v_{2\varphi}. \tag{19.8}$$

Integrating (19.7) twice and substituting (19.8) into the expression obtained, we have

$$f_1 \mu_1 v_{1\varphi} + f_2 \mu_2 v_{2\varphi} = \frac{N}{2} \left(\ln r - \frac{1}{2} \right) r + c_{01} r + \frac{c_{02}}{r}. \tag{19.9}$$

Applying the boundary conditions to (19.9), we determine

$$\left. \begin{array}{c} c_{01} = - \dfrac{N \left[a^2 \left(\ln a - \dfrac{1}{2} \right) - b^2 \left(\ln b - \dfrac{1}{2} \right) \right]}{2 \left(a^2 - b^2 \right)} \\[4mm] c_{02} = \dfrac{N a^2 b^2 \ln \dfrac{a}{b}}{2 \left(a^2 - b^2 \right)} \end{array} \right\} \tag{19.10}$$

Solving (19.8) for $v_{2\varphi}$ and $v_{1\varphi}$ and substituting these expressions in (19.4) and (19.5), we obtain

$$\frac{d^2 v_{1\varphi}}{dr^2} + \frac{1}{r} \frac{dv_{1\varphi}}{dr} - \left(\frac{1}{r^2} + m^2 \right) v_{1\varphi} = Ar + Br \left(\ln r - \frac{1}{2} \right) + \frac{C}{r}, \tag{19.11}$$

$$\frac{d^2 v_{2\varphi}}{dr^2} + \frac{1}{r} \frac{dv_{2\varphi}}{dr} - \left(\frac{1}{r^2} + m^2 \right) v_{2\varphi} = Ar + Br \left(\ln r - \frac{1}{2} \right) + \frac{D}{r}, \tag{19.12}$$

where

$$\left. \begin{array}{ll} A = - \dfrac{K c_{01}}{f_1 \mu_1 f_2 \mu_2}, & C = \dfrac{N}{\mu_1} - \dfrac{K c_{02}}{f_1 \mu_1 f_2 \mu_2} \\[4mm] B = - \dfrac{KN}{2 f_1 \mu_1 f_2 \mu_2}, & D = \dfrac{N}{\mu_2} - \dfrac{K c_{02}}{f_1 \mu_1 f_2 \mu_2} \\[4mm] \multicolumn{2}{c}{m^2 = K \left(\dfrac{1}{f_1 \mu_1} + \dfrac{1}{f_2 \mu_2} \right)} \end{array} \right\} \tag{19.13}$$

Making the substitution R = mr in (19.11), we obtain

$$\frac{d^2 v_{1\varphi}}{dR^2} + \frac{1}{R} \frac{dv_{1\varphi}}{dR} - \left(\frac{1}{R^2} + 1 \right) v_{1\varphi} = \frac{A}{m^3} R + \frac{B}{m^3} R \left(\ln \frac{R}{m} - \frac{1}{2} \right) + \frac{C}{mR}. \tag{19.14}$$

We solve (19.14) by the Lagrange method. It is known that the solution of

$$\frac{d^2 v_{1\varphi}}{dR^2} + \frac{1}{R} \frac{dv_{1\varphi}}{dR} - \left(\frac{1}{R^2} + 1 \right) v_{1\varphi} = 0$$

is

$$v_{1\varphi}^* = c_1 I_1 (R) + c_2 K_1 (R), \tag{19.15}$$

where $I_1(R)$, $K_1(R)$ are the Bessel and Macdonald functions. Considering c_1 and c_2 to be variables, we put

$$c_1' \, I_1(R) + c_2' \, K_1(R) = 0,$$

$$c_1' \, I_1'(R) + c_2' \, K_1'(R) = \frac{A}{m^3} R + \frac{B}{m^3} R \left(\ln \frac{R}{m} - \frac{1}{2} \right) + \frac{C}{mR}.$$

Then

$$\left. \begin{aligned} c_1 &= \int \frac{K_1(R) \left[\dfrac{A}{m^3} R + \dfrac{B}{m^3} R \left(\ln \dfrac{R}{m} - \dfrac{1}{2} \right) + \dfrac{C}{mR} \right] dR}{I_1'(R) K_1(R) - K_1'(R) I_1(R)} \\ c_2 &= \int \frac{I_1(R) \left[\dfrac{A}{m^3} R + \dfrac{B}{m^3} R \left(\ln \dfrac{R}{m} - \dfrac{1}{2} \right) + \dfrac{C}{mR} \right]}{I_1'(R) K_1(R) - K_1'(R) I_1(R)} dR \end{aligned} \right\} \tag{19.16}$$

Let us evaluate the denominator of the integrand in (19.16):

$$I_1'(R) K_1(R) - K_1'(R) I_1(R). \tag{19.17}$$

To do this we utilize equalities which are valid for Bessel functions of purely imaginary argument and for the Macdonald functions:

$$I_0(R) + I_2(R) = 2I_1'(R),$$

$$K_0(R) + K_2(R) = -2K_1'(R).$$

We hence determine $I_1'(R)$, $K_1'(R)$ and, substituting the found values in (19.17), have

$$I_1'(R) K_1(R) - K_1'(R) I_1(R) = \frac{1}{R}.$$

Then

$$\left. \begin{aligned} c_1 &= \int R K_1(R) \left[\frac{A}{m^3} R + \frac{B}{m^3} R \left(\ln \frac{R}{m} - \frac{1}{2} \right) + \frac{C}{mR} \right] dR \\ c_2 &= \int R I_1(R) \left[\frac{A}{m^3} R + \frac{B}{m^3} R \left(\ln \frac{R}{m} - \frac{1}{2} \right) + \frac{C}{mR} \right] dR \end{aligned} \right\}. \tag{19.18}$$

Simple calculation shows that

$$\left. \begin{aligned} \int R^2 K_1(R) \ln R \, dR &= - \left[R^2 K_0(R) + 2R K_1(R) \right] \ln R - R K_1(R) - 2K_0(R) + c_1^0 \\ \int R^2 I_1(R) \ln R \, dR &= \left[R^2 I_0(R) - 2R I_1(R) \right] \ln R - R I_1(R) + 2I_0(R) + c_2^0 \end{aligned} \right\} \tag{19.19}$$

Taking account of the values of the integrals (19.19), we obtain expressions for c_1 and c_2, with whose aid we obtain

$$v_{1\varphi} = \frac{N}{m^2} \left\{ \frac{\dfrac{1}{f_1 \mu_1 + f_2 \mu_2} - \dfrac{1}{\mu_1}}{I_1(ma) K_1(mb) - I_1(mb) K_1(ma)} \left[\frac{I_1(mR) K_1(mb) - K_1(mr) I_1(mb)}{a} - \frac{I_1(mr) K_1(ma) - K_1(mr) I_1(ma)}{b} \right] + \right.$$

$$+ \left(\frac{1}{\mu_1} - \frac{1}{f_1\mu_1 + f_2\mu_2} - \frac{Ka^2 b^2 \ln \frac{a}{b}}{2(a^2 - b^2) f_1\mu_1 f_2\mu_2} \right) \frac{1}{r} - \frac{K}{2 f_1\mu_1 f_2\mu_2} r \left(\ln r - \frac{1}{2} \right) + \frac{K \left[a^2 \left(\ln a - \frac{1}{2} \right) - b^2 \left(\ln b - \frac{1}{2} \right) \right]}{2 f_1\mu_1 f_2\mu_2 (a^2 - b^2)} r \right\}$$

$$\tag{19.20}$$

and analogously

$$v_{2\varphi} = - \frac{N}{m^2} \left\{ \frac{\frac{1}{f_1\mu_1 + f_2\mu_2} - \frac{1}{\mu_2}}{I_1(ma) K_1(mb) - I_1(mb) K_1(ma)} \left[\frac{I_1(mr) K_1(mb) - K_1(mr) I_1(mb)}{a} - \frac{I_1(mr) K_1(ma) - K_1(mr) I_1(ma)}{b} \right] + \right.$$

$$\left. + \left(\frac{1}{\mu_2} - \frac{1}{f_1\mu_1 + f_2\mu_2} - \frac{Ka^2 b^2 \ln \frac{a}{b}}{2(a^2 - b^2) f_1\mu_1 f_2\mu_2} \right) \frac{1}{r} - \frac{K}{2 f_1\mu_1 f_2\mu_2} r \left(\ln r - \frac{1}{2} \right) + \frac{K \left[a^2 \left(\ln a - \frac{1}{2} \right) - b^2 \left(\ln b - \frac{1}{2} \right) \right]}{2 f_1\mu_1 f_2\mu_2 (a^2 - b^2)} r \right\}.$$

$$\tag{19.21}$$

Therefore, we have obtained formulas for the velocity distribution of the first and second media over the cross section of the pipe in the form of (19.20) and (19.21).

The pressure distribution along the radius r is found from the first two equations of the system (19.2). It is readily noted that the equality

$$v_\varphi = - \frac{N}{2\mu} \left[\left(\frac{a^2 \ln a - b^2 \ln b}{a^2 - b^2} - \frac{1}{2} \right) r - r \left(\ln r - \frac{1}{2} \right) - \frac{a^2 b^2 \ln \frac{a}{b}}{a^2 - b^2} \frac{1}{r} \right],$$

which is the formula for the velocity distribution of a viscous incompressible fluid in classical hydrodynamics [10], results from (19.20) and (19.21) for $\mu_1 = \mu_2 = \mu$.

The volume discharges of the media Q_1 and Q_2 are computed from the formulas

$$\left. \begin{array}{l} Q_1 = f_1 \int_b^a v_{1\varphi} \, dr. \\[2mm] Q_2 = f_2 \int_b^a v_{2\varphi} \, dr \end{array} \right\} .$$

$$\tag{19.22}$$

Substituting the expressions for $v_{1\varphi}$, $v_{2\varphi}$ from (19.20) and (19.21) into (19.22) and integrating, we obtain the following formulas for the discharges:

$$Q_1 = - Nf_1 \left\{ \frac{\frac{1}{f_1\mu_1 + f_2\mu_2} - \frac{1}{\mu_1}}{[I_1(ma) K_1(mb) - I_1(mb) K_1(ma)] m^3} \left[\frac{I_1(mb) K_0(ma) + K_1(mb) I_0(ma)}{a} + \right. \right.$$

$$\left. + \frac{I_1(ma) K_0(mb) + K_1(ma) I_0(mb)}{b} - \frac{2}{mab} \right] -$$

$$\left. - \left(\frac{1}{f_1\mu_1 + f_2\mu_2} - \frac{1}{\mu_1} \right) \frac{\ln \frac{a}{b}}{m^2} + \frac{a^2 - b^2}{8(f_1\mu_1 + f_2\mu_2)} - \frac{a^2 b^2 \left(\ln \frac{a}{b} \right)^2}{2(a^2 - b^2)(f_1\mu_1 + f_2\mu_2)} \right\},$$

$$\tag{19.23}$$

$$Q_2 = - Nf_2 \left\{ \frac{\frac{1}{f_1\mu_1 + f_2\mu_2} - \frac{1}{\mu_2}}{[I_1(ma) K_1(mb) - I_1(mb) K_1(ma)] m^3} \left[\frac{I_1(mb) K_0(ma) + K_1(mb) I_0(ma)}{a} + \right. \right.$$

$$\left. + \frac{I_1(ma) K_0(mb) + K_1(ma) I_0(mb)}{b} - \frac{2}{mab} \right] -$$

$$\left. - \left(\frac{1}{f_1\mu_1 + f_2\mu_2} - \frac{1}{\mu_2} \right) \frac{\ln \frac{a}{b}}{m^2} + \frac{a^2 - b^2}{8(f_1\mu_1 + f_2\mu_2)} - \frac{a^2 b^2 \left(\ln \frac{a}{b} \right)^2}{2(a^2 - b^2)(f_1\mu_1 + f_2\mu_2)} \right\}.$$

$$\tag{19.24}$$

In the particular case when $\mu_1 = \mu_2 = \mu$ we have

$$Q = Q_1 + Q_2 = -\frac{N}{8\mu}\left[a^2 - b^2 - \frac{4a^2 b^2}{a^2 - b^2}\left(\ln\frac{a}{b}\right)^2\right],$$

which agrees with the formula for the corresponding problem of the hydrodynamics of a viscous incompressible fluid [10].

§ 20. Discharge Formulas in the Composite Channel Case

Let the channel under consideration consist of three sections (Fig. 6) a rectilinear portion (AB) of length l_1, an arc (BD) with the bend $\varphi = \pi/2$, and the line (DC) of length l_2.

Let the channel thickness h, the entrance pressure p_1 into, and the exit pressure p_4 out of the pipe be known. Let us find the volume discharge formula for the first medium. In the presence of formulas (13.15) and (19.23), the volume discharge becomes

for the section AB

$$Q_1 = \frac{p_1 - p_2}{l_1}f_1 F_1,\tag{20.1}$$

for the section BD

$$Q_1 = \frac{p_2 - p_3}{\frac{\pi}{2}}f_1 F_2,\tag{20.2}$$

for the section DC

$$Q_1 = \frac{p_3 - p_4}{l_2}f_1 F_1,\tag{20.3}$$

where

$$F_1 = 2\left(\frac{\mu_1}{f_1\mu_1 + f_2\mu_2} - 1\right)\left[\frac{f_1 f_2 \mu_2}{K(f_1\mu_1 + f_2\mu_2)}\right]^{\frac{3}{2}}\mu_1^{\frac{1}{2}}\,\mathrm{th}\sqrt{K\frac{f_1\mu_1 + f_2\mu_2}{f_1\mu_1 + f_2\mu_2}\frac{h}{2} + \frac{h^3}{12(f_1\mu_1 + f_2\mu_2)}} -$$
$$-\left(\frac{\mu_1}{f_1\mu_1 + f_2\mu_2} - 1\right)\frac{f_1 f_2\mu_2 h}{K(f_1\mu_1 + f_2\mu_2)};$$

$$F_2 = \frac{\frac{1}{f_1\mu_1 + f_2\mu_2} - \frac{1}{\mu_1}}{[I_1(ma)K_1(mb) - I_1(mb)K_1(ma)]m^3}\left[\frac{I_1(mb)K_0(ma) + K_1(mb)I_0(ma)}{a} + \right.$$
$$\left.+\frac{I_1(ma)K_0(mb) + K_1(ma)I_0(mb)}{b} - \frac{2}{mab}\right] - \left(\frac{1}{f_1\mu_1 + f_2\mu_2} - \frac{1}{\mu_1}\right)\frac{\ln\frac{a}{b}}{m^2} + \frac{a^2 - b^2}{8(f_1\mu_1 + f_2\mu_2)} - \frac{a^2 b^2\left(\ln\frac{a}{b}\right)^2}{2(a^2 - b^2)(f_1\mu_1 + f_2\mu_2)}.$$

It follows from (20.1), (20.2), and (20.3) that

$$p_1 - p_2 = \frac{Q_1 l_1}{f_1 F_1},\quad p_2 - p_3 = \frac{Q_1\pi}{2 f_1 F_2},\quad p_3 - p_4 = \frac{Q_1 l_2}{f_1 F_1}.\tag{20.4}$$

Adding (20.4) term by term, we have

$$p_1 - p_4 = Q_1\left(\frac{l_1 + l_2}{F_1} + \frac{\pi}{2F_2}\right),$$

and hence the expression

$$Q_1 = f_1\frac{p_1 - p_4}{\frac{l_1 + l_2}{F_1} + \frac{\pi}{2F_2}},$$

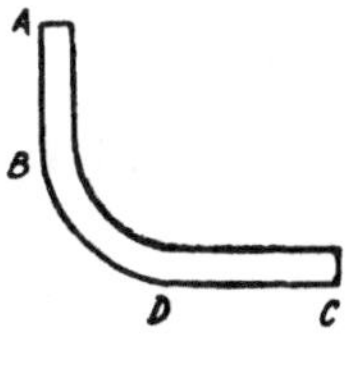

Fig. 6

which is the formula for the volume discharge of the first medium for the composite channel considered. The formula for the discharge of the second medium can be obtained completely analogously.

§ 21. Motion of Two-Phase Viscous Media between Two Parallel Finite Planes*

The influence of conditions at the entrance and exit of a pipe on the motion of two-phase media in conduits will often be quite significant. Conduits for which it is impossible to neglect the conditions at the entrance and exit of the pipe are called finite (short). Leonov [21] solved the problem for an incompressible fluid for such a case.

We represent the differential equations of steady-state motion (6.2), (6.3) as follows:

for the first medium

$$\left. \begin{array}{l} u_1 \dfrac{\partial u_1}{\partial x} + v_1 \dfrac{\partial u_1}{\partial y} = -\dfrac{1}{\rho_{1i}} \dfrac{\partial p}{\partial x} + \nu_1 \Delta u_1 + \dfrac{K}{\rho_1}(u_2 - u_1) \\[3mm] u_1 \dfrac{\partial v_1}{\partial x} + v_1 \dfrac{\partial v_1}{\partial y} = -\dfrac{1}{\rho_{1i}} \dfrac{\partial p}{\partial y} + \nu_1 \Delta v_1 + \dfrac{K}{\rho_1}(v_2 - v_1) \end{array} \right\} \quad (21.1)$$

for the second medium

$$\left. \begin{array}{l} u_2 \dfrac{\partial u_2}{\partial x} + v_2 \dfrac{\partial u_2}{\partial y} = -\dfrac{1}{\rho_{2i}} \dfrac{\partial p}{\partial x} + \nu_2 \Delta u_2 + \dfrac{K}{\rho_2}(u_1 - u_2) \\[3mm] u_2 \dfrac{\partial v_2}{\partial x} + v_2 \dfrac{\partial v_2}{\partial y} = -\dfrac{1}{\rho_{2i}} \dfrac{\partial p}{\partial y} + \nu_2 \Delta v_2 + \dfrac{K}{\rho_2}(v_1 - v_2) \end{array} \right\} \quad (21.2)$$

Let 2h denote the distance between the walls, and L the length of the walls, where $h/L = \delta \ll 1$. The x axis passes through the middle of the pipe in the flow direction. The y axis is perpendicular to the x axis. In order to estimate the members of (21.1) and (21.2), we perform the transformation

$$\left. \begin{array}{l} u_1 = U_1 u, \quad u_2 = U_1 u', \quad v_1 = V_1 v, \quad v_2 = V_1 v' \\[3mm] x = L x_1, \quad y = h y_1, \quad p = \dfrac{\mu_1 v_1 L}{h^2} q \\[3mm] K = \dfrac{\rho_1 V_1}{L} K_1, \quad \mathrm{Re}_1 = \dfrac{U_1 h}{\nu_1} \end{array} \right\} \quad (21.3)$$

where u, u', v, v', q, K_1, x_1, and y_1 are dimensionless quantities; U_1, V_1 the characteristic longitudinal and transverse velocities of the first medium, where $V_1/U_1 = \delta$; and Re_1 is the Reynolds number of the first medium.

Then (21.1) becomes

$$\left. \begin{array}{l} \mathrm{Re}_1 \delta \left(u \dfrac{\partial u}{\partial x_1} + v \dfrac{\partial u}{\partial y_1} \right) = -\dfrac{\partial q}{\partial x_1} + \delta^2 \dfrac{\partial^2 u}{\partial x_1^2} + \dfrac{\partial^2 u}{\partial y_1^2} + \mathrm{Re}_1 \delta K_1 (u' - u) \\[3mm] \mathrm{Re}_1 \delta^3 \left(u \dfrac{\partial v}{\partial x_1} + v \dfrac{\partial v}{\partial y_1} \right) = -\dfrac{\partial q}{\partial y_1} + \delta^2 \dfrac{\partial^2 v}{\partial y_1^2} + \delta^4 \dfrac{\partial^2 v}{\partial x_1^2} + \mathrm{Re}_1 \delta^3 K_1 (v' - v) \end{array} \right\} \quad (21.4)$$

---

*The problems of § 21, 22, 23 were solved by M. P. Nazarii, a graduate student, under the author's supervision.

Let us assume that the dimensionless functions and their derivatives, as well as the Reynolds number, are of the same order of magnitude; then terms containing δ^3, δ^4, and $\delta^2 \, \partial^2 v / \partial y_1^2$ may be discarded. Returning to dimensional variables, we obtain

$$\left. \begin{aligned} u_1 \frac{\partial u_1}{\partial x} + v_1 \frac{\partial u_1}{\partial y} &= -\frac{1}{\rho_{1i}} \frac{\partial p}{\partial x} + \nu_1 \Delta u_1 + \frac{K}{\rho_1}(u_2 - u_1) \\ \frac{\partial p}{\partial y} &= 0, \quad \frac{\partial u_1}{\partial x} + \frac{\partial v_1}{\partial y} = 0 \end{aligned} \right\} \tag{21.5}$$

$$\left. \begin{aligned} u_2 \frac{\partial u_2}{\partial x} + v_2 \frac{\partial u_2}{\partial y} &= -\frac{1}{\rho_{2i}} \frac{\partial p}{\partial x} + \nu_2 \Delta u_2 + \frac{K}{\rho_2}(u_1 - u_2) \\ \frac{\partial p}{\partial y} &= 0, \quad \frac{\partial u_2}{\partial x} + \frac{\partial v_2}{\partial y} = 0 \end{aligned} \right\} \tag{21.6}$$

Linearizing the first equations of the system (21.5) and (21.6) according to Oseen [10], we have

$$\left. \begin{aligned} U_1 \frac{\partial u_1}{\partial x} &= -\frac{1}{\rho_{1i}} \frac{\partial p}{\partial x} + \nu_1 \, \Delta u_1 + \frac{K}{\rho_1}(u_2 - u_1) \\ U_2 \frac{\partial u_2}{\partial x} &= -\frac{1}{\rho_{2i}} \frac{\partial p}{\partial x} + \nu_2 \, \Delta u_2 + \frac{K}{\rho_2}(u_1 - u_2) \end{aligned} \right\} \tag{21.7}$$

where U_1 and U_2 are the average discharge velocities of the appropriate media.

We seek the solution of the system (21.7) in dimensionless form, which affords the possibility of obtaining the final solution in dynamically similar cases of the motion. To this end we replace all the dimensional variables and K in (21.7) by dimensionless quantities:

$$\left. \begin{aligned} u_1 &= U_1 u, \quad v_1 = U_1 v \\ u_2 &= U_1 u', \quad v_2 = U_1 v' \\ p &= \frac{\rho_{1i} U_1^2}{\mathrm{Re}_1} q, \quad K = \frac{\rho_1 U_1}{h \, \mathrm{Re}_1} K' \\ x &= h x_1 \quad y = h y_1 \end{aligned} \right\} \tag{21.8}$$

We then have the system

$$\left. \begin{aligned} \mathrm{Re}_1 \frac{\partial u}{\partial x_1} &= -\frac{\partial q}{\partial x_1} + \Delta u + K'(u' - u) \\ \frac{\partial q}{\partial y} &= 0, \quad \frac{\partial u}{\partial x_1} + \frac{\partial v}{\partial y_1} = 0 \\ \mathrm{Re}_2 \frac{\partial u'}{\partial x_1} &= -\frac{\mu_1}{\mu_2} \frac{\partial q}{\partial x_1} + \Delta u' + \frac{\mu_1 f_1}{\mu_2 f_2} K'(u - u') \\ \frac{\partial u'}{\partial x_1} + \frac{\partial v'}{\partial y'} &= 0 \end{aligned} \right\} \tag{21.9}$$

where the u, u', v, v', q, x_1, y_1, K' are dimensionless quantities; $\mathrm{Re}_2 = U_2 h / \nu_2$ is the Reynolds number of the second medium.

Taking account of the condition of constancy of the discharges

$$\int_0^1 u \, dy_1 = 1, \quad \int_0^1 u' \, dy_1 = U_2 \, U_1, \quad (0 \leqslant x_1 \leqslant l) \tag{21.10}$$

(where $l = L/h$ is the dimensionless length of the conduit) and the second equation of the system (21.9), we have

$$q(x_1) = q_0 + \int \left(\frac{\partial u}{\partial y_1}\right)_1 dx_1 + K'(U_2/U_1 - 1)x_1; \qquad (21.11)$$

where q_0 is a constant of integration.

Substituting (21.11) in the first and third equations of the system (21.9), we obtain

$$\left.\begin{aligned}
\mathrm{Re}_1 \frac{\partial u}{\partial x_1} &= \Delta u + K'(u'-u) - \left(\frac{\partial u}{\partial y_1}\right) - K'\left(\frac{U_2}{U_1} - 1\right) \\
\mathrm{Re}_2 \frac{\partial u'}{\partial x_1} &= \Delta u' + \frac{\mu_1 f_1}{\mu_2 f_2} K'(u-u') - \left(\frac{\partial u'}{\partial y_1}\right)_1 - \frac{\mu_1 f_1}{\mu_2 f_2} K'\left(1 - \frac{U_2}{U_1}\right)
\end{aligned}\right\} \qquad (21.12)$$

where the boundary conditions will now be

$$\left.\begin{aligned}
u_{x_1=0} &= \varphi_1(y_1),\ u|_{x_1=l} = \varphi_2(y_1),\ u|_{y_1=\pm 1} = 0 \\
u'|_{x_1=0} &= \psi_1(y_1),\ u'|_{x_1=l} = \psi_2(y_1),\ u'|_{y=\pm 1} = 0
\end{aligned}\right\} \qquad (21.13)$$

The functions φ_1, φ_2, ψ_1, and ψ_2 should satisfy condition (21.10).

Let us examine the case, analogous to that considered in [21], when it can be assumed that the Reynolds numbers are very small (Re_1, $\mathrm{Re}_2 \ll 1$). The system (21.12) then becomes

$$\left.\begin{aligned}
\Delta u + K'(u'-u) &= \left(\frac{\partial u}{\partial y_1}\right)_1 + K'\left(\frac{U_2}{U_1} - 1\right) \\
\Delta u' + K' \frac{\mu_1 f_1}{\mu_2 f_2}(u-u') &= \left(\frac{\partial u'}{\partial y_1}\right)_1 + \frac{\mu_1 f_1}{\mu_2 f_2} K'\left(1 - \frac{U_2}{U_1}\right)
\end{aligned}\right\} \qquad (21.14)$$

Subtracting the second equation of (21.14) term by term from the first, we obtain

$$\Delta w - a^2 w = \left(\frac{\partial w}{\partial y_1}\right)_1 + a^2\left(\frac{U_2}{U_1} - 1\right); \qquad (21.15)$$

where

$$w = u - u'; \qquad a^2 = \left(1 + \frac{f_1 \mu_1}{f_2 \mu_2}\right) K'.$$

Moreover, multiplying the first equation of (21.14) by $\mu_1 f_1/\mu_2 f_2$ and adding it term by term to the second, we have

$$\Delta w_1 = \left(\frac{\partial w_1}{\partial y_1}\right)_1, \qquad (21.16)$$

where

$$w_1 = \frac{\mu_1 f_1}{\mu_2 f_2} u + u'.$$

The problem is therefore reduced to solving the system of equations (21.15) and (21.16) under the following boundary conditions:

$$w/_{x_1=0} = \varphi_1(y_1) - \psi_1(y_1) = \Theta_1(y_1)$$
$$w_1/_{x_1=l} = \varphi_2(y_1) - \psi_2(y_1) = \Theta_2(y_1)$$
$$w_1/_{x_1=0} = \alpha\varphi_1(y_1) + \psi_1(y_1) = \Omega_1(y_1)$$
$$w_1/_{x_1=l} = \alpha\varphi_2(y_1) + \psi_2(y_1) = \Omega_2(y_1)$$
$$w/_{y_1=\pm 1} = 0$$
$$w_1/_{y_1=\pm 1} = 0$$
$$\alpha = \mu_1 f_1/\mu_2 f_2$$

(21.17)

(for convenience, the subscript 1 is henceforth omitted from x and y).

Let us seek the solution of (21.15) as

$$w(x,y) = \bar{u}(y) + \bar{v}(x,y), \tag{21.18}$$

where the function $\bar{u}(y)$ is determined from

$$\frac{d^2\bar{u}}{dy^2} - a^2\bar{u} = \left(\frac{d\bar{u}}{dy}\right)_1 + a^2\left(\frac{U_2}{U_1} - 1\right) \Bigg\} \tag{21.19}$$
$$\bar{u}/_{y=\pm 1} = 0$$

The general solution of this equation is

$$\bar{u} = c_1 \operatorname{ch}\, ay + c_2 \operatorname{sh}\, ay - \frac{1}{a^2}\left(\frac{d\bar{u}}{dy}\right)_1 - \left(\frac{U_2}{U_1} - 1\right), \tag{21.20}$$

from which

$$\left(\frac{d\bar{u}}{dy}\right)_1 = c_1 a \operatorname{sh}\, a + c_1 a \operatorname{ch}\, a.$$

Substituting this into (21.20), we obtain

$$\bar{u} = c_1\left(\operatorname{ch}\, ay - \frac{1}{a}\operatorname{sh}\, a\right) + c_2\left(\operatorname{sh}\, ay + \frac{1}{a}\operatorname{ch}\, a\right) - \left(\frac{U_2}{U_1} - 1\right).$$

By satisfying the boundary conditions we find the arbitrary constants c_1 and c_2:

$$c_1 = \left(\frac{U_2}{U_1} - 1\right)\Big/\left(\operatorname{ch}\, a - \frac{1}{a}\operatorname{sh}\, a\right) \Bigg\} \tag{21.21}$$
$$c_2 = 0$$

Taking (21.21) into account, we finally write

$$\bar{u}(y) = \frac{\operatorname{ch}\, ay - \operatorname{ch}\, a}{\operatorname{ch}\, a - \frac{1}{a}\operatorname{sh}\, a}\left(\frac{U_2}{U_1} - 1\right). \tag{21.22}$$

The function $\bar{v}(x, y)$ is determined from the homogeneous equation

$$\Delta\bar{v} - a^2\bar{v} - \left(\frac{\partial\bar{v}}{\partial y}\right)_1 = 0, \tag{21.23}$$
$$\bar{v}/_{x=0} = \Theta_1(y) - \bar{u}(y),$$
$$\bar{v}/_{x=l} = \Theta_2(y) - \bar{u}(y), \quad \bar{v}/_{y=\pm 1} = 0.$$

Putting $\bar{v} = X(x)\ Y(y)$ and separating variables, we obtain in place of (21.23)

$$X'' - (\lambda + a^2)\ X = 0, \tag{21.24}$$

$$y'' + \lambda Y = Y'_{(1)}, \quad Y_{y=\pm 1} = 0. \tag{21.25}$$

To the accuracy of an arbitrary factor the solution of (21.25) is

$$Y_n = \cos \beta_n\ y - \cos \beta_n, \tag{21.26}$$

where $\lambda = \beta_n^2$; β_n are the successive roots of the equation $\tan x = x$. Taking account of (21.26) as well as the solution of (21.24), we have

$$v(x,y) = \sum_{n=1}^{\infty} \left(a_n e^{\gamma_n x} + b_n e^{-\gamma_n x} \right) (\cos \beta_n\ y - \cos \beta_n\ ); \tag{21.27}$$

where $\gamma_n^2 = a^2 + \beta_n^2$.

Assuming the functions $\bar{v}\ (0, y)$ and $\bar{v}\ (l, y)$ to be expanded in uniformly and absolutely convergent Fourier series in the functions (21.26), and satisfying the boundary conditions, we find expressions for a_n and b_n from 21.27. Substituting the values found into (21.27), we finally find

$$\bar{v}\ (x,y) = \sum_{n=1}^{\infty} \left[\Theta_n^{(1)} \operatorname{sh} \gamma_n (l - x) + \Theta_n^{(2)} \operatorname{sh} \gamma_n\ x \right] \frac{\cos \beta_n\ y - \cos \beta_n}{\operatorname{sh} \gamma_n\ l}, \tag{21.28}$$

where $\Theta_n^{(i)}(i = 1,2)$ are Fourier coefficients of the form

$$\Theta_n^{(i)} = \frac{2}{\sin^2 \beta_n} \int_0^1 \left[\Theta_i - \frac{\operatorname{ch} ay - \operatorname{ch} a}{\operatorname{ch} a - \frac{1}{a} \operatorname{sh} a} \left(\frac{U_2}{U_1} - 1 \right) \right] (\cos \beta_n\ y - \cos \beta_n). \tag{21.29}$$

After some manipulation we obtain

$$\Theta_n^{(i)} = \frac{2}{\sin^2 \beta_n} \int_0^1 \Theta_i \cos \beta_n\ y\ dy + \frac{2a^2}{\sin \beta_n\ (\beta_n^3 + a^2 \beta_n)} \left(\frac{U_2}{U_1} - 1 \right). \tag{21.30}$$

Taking account of (21.18), (21.22), and (21.28), we have

$$w(x,y) = \frac{\operatorname{ch} ay - \operatorname{ch} a}{\operatorname{ch} a - \frac{1}{a} \operatorname{sh} a} \left(\frac{U_2}{U_1} - 1 \right) + \sum_{n=1}^{\infty} \left[\Theta_n^{(1)} \operatorname{sh} \gamma_n (l - x) + \Theta_n^{(2)} \operatorname{sh} \gamma_n\ x \right] \frac{\cos \beta_n\ y - \cos \beta_n}{\operatorname{sh} \gamma_n\ l}. \tag{21.31}$$

We obtain the solution of (21.16) by an analogous method as

$$w_1(x,y) = \frac{3}{2}\left(\alpha + \frac{U_2}{U_1} \right) (1 - y^2)(A_0 x + B_0) + \sum_{n=1}^{\infty} (A_n e^{-\beta_n x} + B_n e^{-\beta_n x})(\cos \beta_n\ y - \cos \beta_n). \tag{21.32}$$

Utilizing the condition of constancy of the discharges

$$\int_0^1 w_1\ dy = \alpha + \frac{U_2}{U_1},$$

we find that $A_0 = 0$, $B_0 = 1$.

Expanding the functions $\Omega_i(y)$ into Fourier series, we obtain

$$\Omega_i(y)=\frac{3}{2}(1-y^2)\left(\alpha+\frac{U_2}{U_1}\right)+\sum_{n=1}^{\infty}\Omega_n^{(i)}(\cos\beta_n y-\cos\beta_n),\qquad(21.33)$$

$$\Omega_n^{(i)}=\frac{2}{\sin^2\beta_n}\int_0^1\left[\Omega_i-\frac{3}{2}(1-y^2)\left(\alpha+\frac{U_2}{U_1}\right)\right](\cos\beta_n y-\cos\beta_n)\,dy\qquad(21.34)$$

$$(i=1,2).$$

If the equation

$$\int_0^1(1-y^2)\cos\beta_n y\,dy=0$$

is taken into account, the expression for $\Omega_n^{(1)}$ simplifies to

$$\Omega_n^{(i)}=\frac{2}{\sin^2\beta_n}\int_0^1\Omega_i(y)\cos\beta_n y\,dy.\qquad(21.35)$$

We find the arbitrary constants A_n and B_n from (21.17) and the expansion (21.33) by making use of the appropriate boundary conditions. Substituting the values of these constants into (21.32), we have

$$w_1=\frac{3}{2}(1-y^2)\left(\alpha+\frac{U_2}{U_1}\right)+\sum_{n=1}^{\infty}\frac{\Omega_n^{(1)}\mathrm{sh}\beta_n(l-x)+\Omega_n^{(2)}\mathrm{sh}\beta_n x}{\mathrm{sh}\beta_n l}(\cos\beta_n y-\cos\beta_n).\qquad(21.36)$$

We find u and u' by solving the system

$$u-u'=w,$$
$$\alpha u+u'=w_1.$$

Having determined the functions u and u' we find v and v' from the continuity equations, and q from (21.11). Expressions for the dimensionless tangential stresses τ_1 and τ_2 may always be obtained when the velocity formulas are known. The dimensional tangential stresses on the wall are

$$\tau=\tau_1\rho_1 U_1^2/\mathrm{Re}_1,\quad\tau'=\tau_2\rho_2 U_1 U_2/\mathrm{Re}_2,$$

$$u=\frac{3}{2}(1-y^2)\frac{U_1\alpha+U_2}{U_1(1+\alpha)}+\frac{\mathrm{ch}\,\alpha y-\mathrm{ch}\,\alpha}{\mathrm{ch}\,a-\frac{1}{a}\mathrm{sh}\,a}\frac{U_2-U_1}{U_1(1+\alpha)}+\sum_{n=1}^{\infty}\left[\frac{\theta_n^{(1)}\mathrm{sh}\gamma_n(l-x)+\theta_n^{(2)}\mathrm{sh}\gamma_n x}{\mathrm{sh}\gamma_n l}+\right.$$
$$\left.+\frac{\Omega_n^{(1)}\mathrm{sh}\beta_n(l-x)+\Omega_n^{(2)}\mathrm{sh}\beta_n x}{\mathrm{sh}\beta_n l}\right]\frac{\cos\beta_n y-\cos\beta_n}{1+\alpha},$$

$$v=\sum_{n=1}^{\infty}\left[\gamma_n\frac{\theta_n^{(1)}\mathrm{ch}\gamma_n(l-x)-\theta_n^{(2)}\mathrm{ch}\gamma_n x}{\mathrm{sh}\gamma_n l}-\beta_n\frac{\Omega_n^{(2)}\mathrm{ch}\beta_n x-\Omega_n^{(1)}\mathrm{ch}\beta_n(l-x)}{\mathrm{sh}\beta_n l}\right]\frac{\sin\beta_n y-y\sin\beta_n}{\beta_n(1+\alpha)},$$

$$\tau_1=\left(\frac{\partial u}{\partial y}\right)_1=-\frac{3(\alpha U_1+U_2)}{U_1(1+\alpha)}+\frac{a\,\mathrm{ch}\,a}{\mathrm{ch}\,a-\frac{1}{a}\mathrm{sh}\,a}\frac{U_2-U_1}{U_1(1+\alpha)}-\sum_{n=1}^{\infty}\left[\frac{\theta_n^{(1)}\mathrm{sh}\gamma_n(l-x)+\theta_n^{(2)}\mathrm{sh}\gamma_n x}{\mathrm{sh}\gamma_n l}+\right.$$
$$\left.+\frac{\Omega_n^{(1)}\mathrm{sh}\beta_n(l-x)+\Omega_n^{(2)}\mathrm{sh}\beta_n x}{\mathrm{sh}\beta_n l}\right]\frac{\beta_n\sin\beta_n}{1+\alpha},$$

$$u' = \frac{\mathrm{ch}\,ay - \mathrm{ch}\,a}{\mathrm{ch}\,a - \frac{1}{a}\,\mathrm{sh}\,a}\,\frac{U_2 - U_1}{U_1\,(1+\alpha)} - \frac{3}{2}\,\alpha\,(1-y^2)\,\frac{U_1\alpha + U_2}{U_1(1+\alpha)} +$$

$$+ \sum_{n=1}^{\infty}\left[\frac{\Omega_n^{(1)}\mathrm{sh}\,\beta_n\,(l-x) + \Omega_n^{(2)}\mathrm{sh}\,\beta_n\,x}{\mathrm{sh}\,\beta_n\,l} - \alpha\,\frac{\Theta_n^{(1)}\mathrm{sh}\gamma_n\,(l-x) + \Theta_n^{(2)}\mathrm{sh}\gamma_n\,x}{\mathrm{sh}\gamma_n\,l}\right]\frac{\cos\beta_n\,y - \cos\beta_n}{1+\alpha},$$

$$v' = \sum_{n=1}^{\infty}\left[\beta_n\,\frac{\Omega_n^{(1)}\mathrm{ch}\beta_n\,(l-x) - \Omega_n^{(2)}\mathrm{ch}\beta_n\,x}{\mathrm{ch}\beta_n\,l} + \alpha\gamma_n\,\frac{\Theta_n^{(2)}\mathrm{ch}\gamma_n\,x - \Theta_n^{(1)}\mathrm{ch}\gamma_n\,(l-x)}{\mathrm{sh}\gamma_n\,l}\right]\frac{\sin\beta_n\,y - y\sin\beta_n}{\beta_n\,(1+\alpha)},$$

$$\tau_2 = \left(\frac{\partial u'}{\partial y}\right)_1 = 3\alpha\,\frac{\alpha U_1 + U_2}{U_1(1+\alpha)} + \frac{a\,\mathrm{sh}\,a}{\mathrm{ch}\,a - \frac{1}{a}\,\mathrm{sh}\,a}\,\frac{U_2 - U_1}{U_1\,(1+\alpha)} - \sum_{n=1}^{\infty}\left[\frac{\Omega_n^{(1)}\mathrm{sh}\beta_n\,(l-x) + \Omega_n^{(2)}\mathrm{sh}\,\beta_n\,x}{\mathrm{sh}\,\beta_n l} - \right.$$

$$\left. - \alpha\,\frac{\Theta_n^{(1)}\mathrm{sh}\gamma_n\,(l-x) + \Theta_n^{(2)}\mathrm{sh}\,\gamma_n\,x}{\mathrm{sh}\,\gamma_n l}\right]\frac{\beta_n\sin\beta_n}{1+\alpha};$$

$$q = q_0 - 3x\,\frac{U_1\alpha + U_2}{U_1(1+\alpha)} + ax\,\frac{\mathrm{sh}\,a}{\mathrm{ch}\,a - \frac{1}{a}\,\mathrm{sh}\,a}\,\frac{U_2 - U_1}{U_1(1+\alpha)} - \sum_{n=1}^{\infty}\left[\frac{\Theta_n^{(2)}\mathrm{ch}\gamma_n\,x - \Theta_n^{(1)}\mathrm{ch}\gamma_n\,(l-x)}{\gamma_n\,\mathrm{sh}\,\gamma_n\,l} + \right.$$

$$\left. + \frac{\Omega_n^{(2)}\mathrm{ch}\beta_n\,x - \Omega_n^{(1)}\mathrm{ch}\beta_n\,(l-x)}{\beta_n\,\mathrm{sh}\beta_n\,l}\right]\frac{\beta_n\sin\beta_n}{1+\alpha} + K'\left(\frac{U_2}{U_1} - 1\right)x.$$

The formulas obtained above express the distributions of the velocity components and the pressure over the pipe section, and also the tangential stresses on the wall, for viscous two-phase media in combined mutually penetrating motion between two parallel plane finite walls at very low Reynolds numbers. Having evaluated the values of the dimensionless variables from these formulas, we easily find the values of the appropriate dimensional quantities from (21.8).

If

$$\mu_1 = \mu_2, \quad \rho_{1i} = \rho_{2i}$$

(i.e., $\varphi_1 = \psi_1$), then the velocity formulas become [21]

$$u = u' = \frac{3}{2}(1-y^2) + \sum_{n=1}^{\infty}\frac{\varphi_n^{(1)}\mathrm{sh}\beta_n\,(l-x) + \varphi_n^{(2)}\mathrm{sh}\beta_n\,x}{\mathrm{sh}\beta_n\,l}\,(\cos\beta_n\,y - \cos\beta_n).$$

§ 22. Motion of Viscous Two-Phase Media in a Finite Circular Cylindrical Conduit

Let us elucidate the solution of the problem in the preceding section for axisymmetric flow of incompressible viscous two-phase media in a cylindrical conduit [22]. Let the z axis be directed along the cylinder axis, and let R denote the radius of the conduit, L its length, and $R/L = \delta \ll 1$.

Let us neglect mass forces in the equations of motion. Let us introduce dimensionless variables (as in the preceding section) and the Reynolds number by putting

$$v_{1z} = U_1 u_1, \quad v_{2z} = U_1 u_2, \quad v_{1r} = V_1 v_1,$$

$$v_{2r} = V_1 v_2, \quad p = \frac{\mu_1 U_1 L}{R^2}p_0, \quad K = \frac{\rho_1 U_1 L}{L}K_0,$$

$$z = L\xi, \quad r = R\eta, \quad \mathrm{Re}_1 = U_1 R/\nu_1,$$

where u_1, u_2, v_1, v_2, p_0, K_0, ξ, η are corresponding dimensionless quantities. We estimate the terms in the resulting system of dimensionless equations, and again returning to the dimensional quantities, obtain the following system of differential equations in a cylindrical coordinate system:

$$
\left.
\begin{aligned}
v_{1z}\frac{\partial v_{1z}}{\partial z}+v_{1r}\frac{\partial v_{1z}}{\partial r}&=-\frac{1}{\rho_{1i}}\frac{\partial p}{\partial z}+\nu_1\Delta v_{1z}+\frac{K}{\rho_1}(v_{2z}-v_{1z})\\[2mm]
\frac{\partial p}{\partial r}&=0,\ \frac{\partial v_{1z}}{\partial z}+\frac{1}{r}\frac{\partial}{\partial r}(rv_{1r})=0
\end{aligned}
\right\}
\tag{22.1}
$$

and

$$
\left.
\begin{aligned}
v_{2z}\frac{\partial v_{2z}}{\partial z}+v_{2r}\frac{\partial v_{2z}}{\partial r}&=-\frac{1}{\rho_{2i}}\frac{\partial p}{\partial z}+\nu_2\Delta v_{2z}+\frac{K}{\rho_2}(v_{1z}-v_{2z})\\[2mm]
\frac{\partial v_{2z}}{\partial z}&+\frac{1}{r}\frac{\partial}{\partial r}(rv_{2r})=0
\end{aligned}
\right\}
\tag{22.2}
$$

$$
\left(\ \Delta=\frac{\partial^2}{\partial z^2}+\frac{1}{r}\frac{\partial}{\partial r}\,r\,\frac{\partial}{\partial r}\right).
$$

Linearizing the first equations in the systems (22.1), (22.2) according to Oseen, and inserting the dimensionless variables and Reynolds number,

$$
\left.
\begin{aligned}
v_{1z}&=U_1u_1,\quad v_{2z}=U_1u_2,\quad v_{1r}=U_1v_1,\quad v_{2r}=U_1v_2\\[2mm]
r&=R\eta,\ z=R\xi,\ p=\frac{\rho_{1i}U_1^2}{\mathrm{Re}_1}q,\ K=\frac{\rho_1 U_1}{R\,\mathrm{Re}_1}K'\\[2mm]
\mathrm{Re}_1&=U_1R/\nu_1,\quad \mathrm{Re}_2=U_2R/\nu_2
\end{aligned}
\right\},
\tag{22.3}
$$

where U_1 and U_2 are the average discharge velocities of the appropriate media, we obtain

$$
\left.
\begin{aligned}
\mathrm{Re}_1\frac{\partial u_1}{\partial \xi}&=-\frac{\partial q}{\partial \xi}+\Delta u_1+K'(u_2-u_1)\\[2mm]
\frac{\partial q}{\partial \eta}&=0,\ \frac{\partial u_1}{\partial \xi}+\frac{1}{\eta}\frac{\partial}{\partial \eta}(\eta v_1)=0\\[2mm]
\mathrm{Re}_2\frac{\partial u_2}{\partial \xi}&=-\frac{\mu_1}{\mu_2}\frac{\partial q}{\partial \xi}+\Delta u_2+\frac{\mu_1 f_1}{\mu_2 f_2}K'(u_1-u_2)\\[2mm]
\frac{\partial u_2}{\partial \xi}&+\frac{1}{\eta}\frac{\partial}{\partial \eta}(\eta v_2)=0
\end{aligned}
\right\}.
\tag{22.4}
$$

Using the conditions of constancy of the discharges in the form

$$
\left.
\begin{aligned}
2\int_0^1 \eta u_1(\xi,\eta)\,d\eta&=1\\[2mm]
2\int_0^1 \eta u_2(\xi,\eta)\,d\eta&=\frac{U_2}{U_1}
\end{aligned}
\right\},
\tag{22.5}
$$

we find

$$
\frac{\partial q}{\partial \xi}=2\left(\frac{\partial u_1}{\partial \eta}\right)_1+K'\left(\frac{U_2}{U_1}-1\right).
\tag{22.6}
$$

Substituting (22.6) into (22.4), we obtain elliptic equations of the form

$$\left.\begin{aligned}
\mathrm{Re}_1 \frac{\partial u_1}{\partial \xi} &= \Delta u_1 + K'(u_2 - u_1) - 2\left(\frac{\partial u_1}{\partial \eta}\right)_1 - K'\left(\frac{U_2}{U_1} - 1\right) \\
\mathrm{Re}_2 \frac{\partial u_2}{\partial \xi} &= \Delta u_2 + \frac{\mu_1 f_1}{\mu_2 f_2} K'(u_1 - u_2) - 2\left(\frac{\partial u_2}{\partial \eta}\right)_1 - \frac{\mu_1 f_1}{\mu_2 f_2} K'\left(1 - \frac{U_2}{U_1}\right)
\end{aligned}\right\} \tag{22.7}$$

for the fundamental velocities of the media. The boundary conditions will be $(l = L/R)$:

$$\left.\begin{aligned}
u_1/_{\xi=0} &= \varphi_1(\eta), \ \ u_1/_{\xi=l} = \varphi_2(\eta), \ \ u_1/_{\eta=1} = 0, \ \ u_1/_{\eta=0} < \infty \\
u_2/_{\xi=0} &= \psi_1(\eta), \ \ u_2/_{\xi=l} = \psi_2(\eta), \ \ u_2/_{\eta=1} = 0, \ \ u_2/_{\eta=0} < \infty
\end{aligned}\right\}, \tag{22.8}$$

where the functions $\varphi_1, \ \varphi_2, \psi_1, \ \psi_2$ satisfy conditions (22.5).

The problem of determining the velocities in two-phase viscous fluid motion in a circular cylindrical conduit has therefore been reduced to solving the system of differential equations (22.7) with the boundary conditions (22.8).

Now, let the Reynolds numbers of the media differ slightly, i.e., $\mathrm{Re}_1 \approx \mathrm{Re}_2 = \mathrm{Re}$. Subtracting the second equation of (22.7) from the first, we have

$$\mathrm{Re}\,\frac{\partial w_1}{\partial \xi} - \Delta w_1 + A w_1 + 2\left(\frac{\partial w_1}{\partial \eta}\right)_1 + B = 0; \tag{22.9}$$

where

$$w_1 = u_1 - u_2, \quad A = \left(1 - \frac{\mu_1 f_1}{\mu_2 f_2}\right) K',$$

$$B = A\left(\frac{U_2}{U_1} - 1\right).$$

The boundary conditions (22.8) and (22.5) become

$$\left.\begin{aligned}
w_1/_{\xi=0} &= \varphi_1 - \psi_1 = \Theta_1(\eta) \\
w_1/_{\xi=l} &= \varphi_2 - \psi_2 = \Theta_2(\eta) \\
w_1/_{\eta=1} &= 0; \quad w_1/_{\eta=0} < \infty \\
2\int_0^1 \eta w_1(\xi,\eta)\,d\eta &= 1 - \frac{U_2}{U_1}
\end{aligned}\right\}. \tag{22.10}$$

We seek the solution of (22.9) as

$$w_1(\xi,\eta) = u(\eta) + v(\xi,\eta), \tag{22.11}$$

where the function $u(\eta)$ is determined from the equation

$$\frac{d^2 u}{d\eta^2} + \frac{1}{\eta}\frac{du}{d\eta} - Au = 2\left(\frac{du}{d\eta}\right)_1 + B \tag{22.12}$$

with the boundary conditions

$$u/_{\eta=1} = 0, \quad u/_{\eta=0} < \infty. \tag{22.13}$$

The general solution of (22.12) will be

$$u = c_1 I_0(\sqrt{A}\,\eta) + c_2 K_0(\sqrt{A}\,\eta) - \frac{2}{A}\left(\frac{du}{d\eta}\right)_1 - \frac{B}{A}. \tag{22.14}$$

Since $K_0(0) = \infty$ and the velocity on the conduit axis should be finite, the constant c_2 must be equated to zero; then c_1 is determined by utilizing the boundary conditions (22.13):

$$u(\eta) = \frac{B}{A}\frac{I_0(\sqrt{A}\,\eta) - I_0(\sqrt{A})}{I_2(\sqrt{A})}. \tag{22.15}$$

The function $v(\xi,\eta)$ is found from the homogeneous equation

$$\mathrm{Re}\frac{\partial v}{\partial \xi} - \Delta v + Av + 2\left(\frac{\partial v}{\partial \eta}\right)_1 = 0 \tag{22.16}$$

with the boundary conditions

$$\left.\begin{array}{c} v/_{\xi=0} = \Theta_1(\eta) - u(\eta), \quad v/_{\xi=0} = \Theta_2(\eta) - u(\eta) \\ v/_{\eta=1} = 0, \quad v/_{\eta=0} < \infty \end{array}\right\}. \tag{22.17}$$

We solve (22.16) by separation of variables. Putting $v = X(\xi)\,Y(\eta)$ and substituting into (22.16), we obtain

$$X'' - \mathrm{Re}\,X' - (\lambda + A)X = 0, \tag{22.18}$$

$$Y'' + \frac{1}{\eta}Y' + \lambda Y = 2Y'(1), \quad |Y(0)| < \infty, \quad Y(1) = 0. \tag{22.19}$$

The general solution of (22.19) is

$$Y = c_3 J_0(\sqrt{\lambda}\,\eta) + c_4 Y_0(\sqrt{\lambda}\,\eta) + \frac{2y'(1)}{\lambda}. \tag{22.20}$$

From the condition of boundedness of the axial velocity we determine $c_4 = 0$. Finding the value of $Y'(1)$ from (22.20), we have for Y

$$Y = c_3\left[J_0(\sqrt{\lambda}\,\eta) + \frac{2}{\sqrt{\lambda}}J_0'(\sqrt{\lambda})\right]. \tag{22.21}$$

Utilizing the recursion relations

$$J_0'(x) = -J_1(x), \quad \frac{2}{\sqrt{\lambda}}J_1(x) = J_0(x) + J_2(x)$$

and the boundary condition $v(\xi,1) = 0$, we find the eigenvalues of (22.19) from

$$J_2(x) = 0. \tag{22.22}$$

If the roots of this equation are denoted by β_k, then solution (22.21), correct to an arbitrary factor, will be

$$Y_k = J_0(\beta_k\,\eta) - J_0(\beta_k). \tag{22.23}$$

By virtue of (22.18) and (22.23), the general solution of (22.16) is written

$$v(\xi,\eta) = \sum_{k=1}^{\infty} (A_k e^{m_k x} + B_k e^{-n_k x})\left[J_0(\beta_k\,\eta) - J_0(\beta_k)\right], \tag{22.24}$$

where

$$m_k = \mathrm{Re}/2 + \sqrt{\mathrm{Re}^2/4 + \beta_k^2 + A} \Bigg\}$$
$$n_k = -\,\mathrm{Re}/2 + \sqrt{\mathrm{Re}^2/4 + \beta_k^2 + A} \Bigg\} \tag{22.25}$$

Expanding $v(0,\eta)$ and $v(l,\eta)$ in Fourier series in the functions (22.23), we obtain ($i=1,2$)

$$\Theta_i = \frac{B}{A}\,\frac{I_0(\sqrt{A}\,\eta) - I_0(\sqrt{A})}{I_2(\sqrt{A})} + \sum_{k=1}^{\infty} \Theta_k^{(l)}\big[J_0(\beta_k\,\eta) - J_0(\beta_k)\big]$$
$$\Theta_k^{(i)} = \frac{2}{J_1^2(\beta_k)}\int_0^1 \eta\left[\Theta_l - \frac{B}{A}\,\frac{I_0(\sqrt{A}\,\eta) - I_0(\sqrt{A})}{I_2(\sqrt{A})}\right]\big[J_0(\beta_k\,\eta) - J_0(\beta_k)\big]\,d\eta \tag{22.26}$$

After some manipulation, the expression for the Fourier coefficients $\Theta_k^{(i)}$ becomes

$$\Theta_k^{(l)} = \frac{2}{J_1^2(\beta_k)}\int_0^1 \eta\Theta_l\,J_0(\beta_k\,\eta)\,d\eta + \frac{A\eta_0(\beta_k)}{J_1^2(\beta_k)(\beta_k^2 + A)}\left(\frac{U_2}{U_1} - 1\right). \tag{22.27}$$

We find A_k and B_k by using the expansion (22.26) and the boundary conditions (22.17). Substituting the values found for A_k and B_k into (22.24), we have

$$v(\xi,\eta) = \sum_{k=1}^{\infty}\Big[\Theta_k^{(1)}\big(e^{m_k l - n_k \xi} - e^{m_k \xi - n_k l}\big) + \Theta_k^{(2)}\big(e^{m_k \xi} - e^{n_k \xi}\big)\Big]\frac{J_0(\beta_k\eta) - J_0(\beta_k)}{e^{m_k \eta} - e^{-n_k l}}.$$

The final solution of (22.9) is obtained as

$$w_1(\xi,\eta) = \frac{I_0(\sqrt{A}\,\eta) - I_0(\sqrt{A})}{I_2(\sqrt{A})}\left(\frac{U_2}{U_1} - 1\right) + \sum_{k=1}^{\infty}\Big[\Theta_k^{(1)}\big(e^{m_k l - n_k \xi} - e^{m_k \xi - n_k l}\big) + \Theta_k^{(2)}\big(e^{m_k \xi} - e^{-n_k \xi}\big)\Big]\frac{J_0(\beta_k\eta) - J_0(\beta_k)}{e^{m_k l} - e^{-n_k l}}. \tag{22.28}$$

Now, multiplying the first equation of the system (22.7) by $f_1\mu_1/f_2\mu_2$ and adding it to the second equation, we write

$$\mathrm{Re}\,\frac{\partial w_2}{\partial \xi} = \Delta w_2 - 2\left(\frac{\partial w_2}{\partial \eta}\right)_1, \tag{22.29}$$

where

$$w_2(\xi,\eta) = \frac{f_1\mu_1}{f_2\mu_2}u_1 + u_2. \tag{22.30}$$

The boundary conditions (22.8) and (22.5) become

$$w_2\big|_{\xi=0} = \alpha\varphi_1 + \psi_1 = \mathcal{Q}_1(\eta) \Bigg\}$$
$$w_2\big|_{\xi=l} = \alpha\varphi_2 + \psi_2 = \mathcal{Q}_2(\eta)$$
$$w_2\big|_{\eta=0} = 0, \quad w_2\big|_{\eta=0} < \infty$$
$$2\int_0^1 \eta w_2(\xi,\eta)\,d\eta = \alpha + \frac{U_2}{U_1} \Bigg\} \tag{22.31}$$

where $\alpha = \mu_1 f_1/\mu_2 f_2$.

We solve (22.29) also by separation of variables. Putting $w_2 = X(\xi) \, Z(\eta)$ and substituting it into (22.29), we find

$$X'' - \text{Re} \, X' - \lambda X = 0, \tag{22.32}$$

$$Z'' + \frac{1}{\eta} Z' + \lambda Z' = 2Z'(1), \tag{22.33}$$

$$|Z(0)| < \infty, \quad Z(1) = 0.$$

Let us note that $\lambda = 0$ is an eigenvalue of (22.33); the corresponding eigenfunction is

$$Z_0(\eta) = 2\left(\alpha + \frac{U_2}{U_1}\right)(1 - \eta^2). \tag{22.34}$$

The arbitrary factor in (22.34) is selected so that the last condition in (22.5) is satisfied. For nonzero eigennumbers, the eigenfunction is expressed (to the accuracy of a constant factor) as

$$\left.\begin{array}{c} \lambda = \beta_k^2 \\[4pt] Z_k(\eta) = J_0(\beta_k \eta) - J_0(\beta_k) \\[4pt] (k = 1, 2, \ldots) \end{array}\right\} \tag{22.35}$$

where β_k are successive roots of the equation

$$J_2(x) = 0.$$

We then write the general solution as

$$w_2(\xi, \eta) = 2\left(\alpha + \frac{U_2}{U_1}\right)(1 - \eta^2) + \sum_{k=1}^{\infty}\left[\Omega_k^{(1)}\left(e^{\alpha_k l - \gamma_k \xi} - e^{\alpha_k \xi - \gamma_k l}\right) + \Omega_k^{(2)}\left(e^{\alpha_k \xi} - e^{-\gamma_k \xi}\right)\right] \frac{J_0(\beta_k \eta) - J_0(\beta_k)}{e^{\alpha_k \xi} - e^{-\gamma_k l}};$$

where

$$\left.\begin{array}{c} \alpha_k = \text{Re}/2 + \sqrt{\text{Re}^2/4 + \beta_k^2} \\[4pt] \gamma_k = -\text{Re}/2 + \sqrt{\text{Re}^2/4 + \beta_k^2} \end{array}\right\} \tag{22.36}$$

and $\Omega_n^{(i)}$ are Fourier coefficients of the form

$$\Omega_k^{(l)} = \frac{2}{J_1^2(\beta_k)} \int_0^1 \eta\left[\Omega_l - 2\left(\alpha + \frac{U_2}{U_1}\right)(1 - \eta^2)\right]\left[J_0(\beta_k \eta) - J_0(\beta_k)\right] d\eta.$$

Taking account of the equality

$$\int_0^1 \eta(1 - \eta^2) J_0(\beta_k \eta) \, d\eta = 0,$$

we have

$$\Omega_k^{(i)} = \frac{2}{J_1^2(\beta_k)} \int\limits_0^1 \eta \Omega_i(\eta) J_0(\beta_k \eta) \, d\eta. \qquad (22.37)$$

Since

$$w_1 = u_1 - u_2,$$
$$w_2 = \alpha u_1 + u_2,$$

we obtain, by solving these equations for u_1 and u_2;

$$u_1 = \frac{2}{1+\alpha}\left(\alpha + \frac{U_2}{U_1}\right)(1-\eta^2) + \frac{I_0(\sqrt{A}\eta)-I_0(\sqrt{A})}{I_2(\sqrt{A})(1+\alpha)}\left(\frac{U_2}{U_1}-1\right) + \sum_{k=1}^{\infty}\left[\frac{\theta_k^{(1)}\left(e^{m_k l - n_k \xi}-e^{m_k \xi - n_k l}\right)+\theta_k^{(2)}\left(e^{m_k \xi}-e^{-n_k \xi}\right)}{e^{m_k l}-e^{-n_k l}} + \right.$$
$$\left. + \frac{\Omega_k^{(1)}\left(e^{\alpha_k l - \gamma_k \xi}-e^{\alpha_k \xi - \gamma_k l}\right)+\Omega_k^{(2)}\left(e^{\alpha_k \xi}-e^{-\gamma_k \xi}\right)}{e^{\alpha_k l}-e^{-\gamma_k l}}\right]\frac{J_0(\beta_k \eta)-J_0(\beta_k)}{1+\alpha},$$

$$u_2 = \frac{2}{1+\alpha}\left(\alpha + \frac{U_2}{U_1}\right)(1-\eta^2) - \frac{\alpha}{1+\alpha}\frac{I_0(\sqrt{A}\eta)-I_0(\sqrt{A})}{I_2(\sqrt{A})}\left(\frac{U_2}{U_1}-1\right) + \sum_{k=1}^{\infty}\left[\frac{\Omega_k^{(1)}\left(e^{\alpha_k l - \gamma_k \xi}-e^{\alpha_k \xi - \gamma_k l}\right)+\Omega_k^{(2)}\left(e^{\alpha_k \xi}-e^{-\gamma_k \xi}\right)}{e^{\alpha_k l}-e^{-\gamma_k l}} - \right.$$
$$\left. - \frac{\alpha\left[\theta_k^{(1)}\left(e^{m_k l - n_k \xi}-e^{m_k \xi - n_k l}\right)+\theta_k^{(2)}\left(e^{m_k \xi}-e^{-n_k \xi}\right)\right]}{e^{m_k l}-e^{-n_k l}}\right]\frac{J_0(\beta_k \eta)-J_0(\beta_k)}{1+\alpha}.$$

We find expressions for the transverse velocity components v_1 and v_2 from the continuity equation (22.4):

$$v_1 = \sum_{k=1}^{\infty}\left[\frac{m_k e^{m_k \xi}\left(\theta_k^{(1)} e^{-n_k l}-\theta_k^{(2)}\right)-n_k e^{-n_k \xi}\left(\theta_k^{(2)}-\theta_k^{(1)} e^{m_k l}\right)}{e^{m_k l}-e^{-n_k l}} + \frac{\alpha_k e^{\alpha_k \xi}\left(\Omega_k^{(1)} e^{-\gamma_k l}-\Omega_k^{(2)}\right)-\gamma_k e^{-\gamma_k \xi}\left(\Omega_k^{(2)}-\Omega_k^{(1)} e^{\alpha_k l}\right)}{e^{\alpha_k l}-e^{-\gamma_k l}}\right]\frac{J_1(\beta_k \eta)-\eta J_1(\beta_k)}{\beta_k(1+\alpha)},$$

$$v_2 = \sum_{k=1}^{\infty}\left[\frac{\alpha_k e^{\alpha_k \xi}\left(\Omega_k^{(1)} e^{-\gamma_k l}-\Omega_k^{(2)}\right)-\gamma_k e^{-\gamma_k \xi}\left(\Omega_k^{(2)}-\Omega_k^{(1)} e^{\alpha_k l}\right)}{e^{\alpha_k l}-e^{-\gamma_k l}} - \right.$$
$$\left. - \frac{\alpha\left[m_k e^{m_k \xi}\left(\theta_k^{(1)} e^{-n_k l}-\theta_k^{(2)}\right)-n_k e^{-n_k \xi}\left(\theta_k^{(2)}-\theta_k^{(1)} e^{m_k l}\right)\right]}{e^{m_k l}-e^{-n_k l}}\right]\frac{J_1(\beta_k \eta)-\eta J_1(\beta_k)}{\beta_k(1+\alpha)}.$$

We find the pressure from (22.6):

$$q = q_0 - \frac{8}{1+\alpha}\left(\alpha + \frac{U_2}{U_1}\right)\xi + 2\sqrt{A}\,\xi\,\frac{I_1(\sqrt{A})}{I_2(\sqrt{A})(1+\alpha)}\left(\frac{U_2}{U_1}-1\right) + 2\sum_{k=1}^{\infty}\left[\frac{m_k e^{-n_k \xi}\left(\theta_k^{(1)} e^{m_k l}-\theta_k^{(2)}\right)-n_k e^{m_k \xi}\left(\theta_k^{(2)}-\theta_k^{(1)} e^{-n_k l}\right)}{(A+\beta_k^2)\left(e^{m_k l}-e^{-n_k l}\right)} + \right.$$
$$\left. + \frac{\alpha_k e^{-\gamma_k \xi}\left(\Omega_k^{(1)} e^{\alpha_k l}-\Omega_k^{(2)}\right)-\gamma_k e^{\alpha_k \xi}\left(\Omega_k^{(2)}-\Omega_k^{(1)} e^{-\gamma_k l}\right)}{\beta_k^2\left(e^{\alpha_k l}-e^{-\gamma_k l}\right)}\right]\frac{\beta_k J_1(\beta_k)}{1+\alpha} + K'\left(\frac{U_2}{U_1}-1\right)\xi.$$

Knowing the values of the longitudinal velocities u_1 and u_2, we easily determine the dimensionless tangential stresses τ, τ' on the wall:

$$\tau = \left(\frac{\partial u'}{\partial \eta}\right)_1 = -\frac{4}{1+\alpha}\left(\alpha + \frac{U_2}{U_1}\right) + \frac{\sqrt{A}\,I_1(\sqrt{A})}{I_2(\sqrt{A})(1+\alpha)}\left(\frac{U_2}{U_1}-1\right) -$$

$$-\sum_{k=1}^{\infty}\left[\frac{\theta_k^{(1)}\left(e^{m_kl-n_k\xi}-e^{m_k\xi-n_kl}\right)+\theta_k^{(2)}\left(e^{m_k\xi}-e^{-n_k\xi}\right)}{e^{m_kl}-e^{-n_kl}}+\frac{\Omega_k^{(1)}\left(e^{\alpha_kl-\gamma_k\xi}-e^{\alpha_k\xi-\gamma_kl}\right)+\Omega_k^{(2)}\left(e^{\alpha_k\xi}-e^{-\gamma_k\xi}\right)}{e^{\alpha_kl}-e^{-\gamma_kl}}\right]\frac{\beta_kJ_1(\beta_k)}{1+\alpha},$$

$$\tau'=\left(\frac{\partial u_2}{\partial\eta}\right)_1=-\frac{4}{1+\alpha}\left(\alpha+\frac{U_2}{U_1}\right)-\frac{\alpha}{1+\alpha}\frac{\sqrt{A}\,I_1(\sqrt{A})}{I_2(\sqrt{A})}\left(\frac{U_2}{U_1}-1\right)-$$

$$-\sum_{k=1}^{\infty}\left[\frac{\Omega_k^{(1)}\left(e^{\alpha_kl-\gamma_k\xi}-e^{\alpha_k\xi-\gamma_kl}\right)+\Omega_k^{(2)}\left(e^{\alpha_k\xi}-e^{-\gamma_k\xi}\right)}{e^{\alpha_kl}-e^{-\gamma_kl}}-\frac{\alpha\left[\theta_k^{(1)}\left(e^{m_kl-n_k\xi}-e^{m_k\xi-n_kl}\right)+\theta_k^{(2)}\left(e^{m_k\xi}-e^{-n_k\xi}\right)\right]}{e^{m_kl}-e^{-n_kl}}\right]\frac{\beta_kJ_1(\beta_k)}{1+\alpha};$$

the dimensional tangential stresses on the wall are

$$\tau_1=\tau\rho_1\,U_1^2/\mathrm{Re}_1,$$

$$\tau_2=\tau'\rho_2\,U_1\,U_2/\mathrm{Re}_2.$$

In the case $\mu_1=\mu_2$, $\rho_{1i}=\rho_{2i}$ (i.e., $\varphi_1=\psi_1$) the solutions obtained agree with the results for single-phase motion [21]:

$$u=u_1=u_2=2(1-\eta^2)+\sum_{k=1}^{\infty}\left[\varphi_k^1\left(e^{\alpha_kl-\gamma_k\xi}-e^{\alpha_k\xi-\gamma_kl}\right)+\varphi_k^{(2)}\left(e^{\alpha_k\xi}-e^{-\gamma_k\xi}\right)\right]\frac{J_0(\beta_k\eta)-J_0(\beta_k)}{e^{\alpha_kl}-e^{-\gamma_kl}}.$$

If the Reynolds numbers are quite small ($\mathrm{Re}_1\ll1$, $\mathrm{Re}_2\ll1$), then by passing to the limit as $\mathrm{Re}\to0$ in the previous formulas, we obtain the solution of the system (22.7). Here

$$m_k,\quad n_k\to\sqrt{A+\beta_k^2},$$

$$\alpha_k,\quad \gamma_k\to\beta_k.$$

In this case we obtain for the longitudinal velocities

$$u_1=\frac{2}{1+\alpha}\left(\alpha+\frac{U_2}{U_1}\right)(1-\eta^2)+\frac{I_0(\sqrt{A}\,\eta)-I_0(\sqrt{A})}{I_2(\sqrt{A})(1+\alpha)}\left(\frac{U_2}{U_1}-1\right)+\sum_{k=1}^{\infty}\left[\frac{\theta_k^{(1)}\mathrm{sh}\sqrt{A+\beta_k^2}\,(l-\xi)+\theta_k^{(2)}\,\mathrm{sh}\sqrt{A+\beta_k^2}\,\xi}{\mathrm{sh}\sqrt{A+\beta_k^2}\,l}+\right.$$

$$\left.+\frac{\Omega_k^{(1)}\,\mathrm{sh}\,\beta_k(l-\xi)+\Omega_k^{(2)}\,\mathrm{sh}\,\beta_k\xi}{\mathrm{sh}\,\beta_k\,l}\right]\frac{J_0(\beta_k\eta)-J_0(\beta_k)}{1+\alpha},$$

$$u_2=\frac{2}{1+\alpha}\left(\alpha+\frac{U_2}{U_1}\right)(1-\eta^2)-\frac{\alpha}{1+\alpha}\frac{I_0(\sqrt{A}\,\eta)-I_0(\sqrt{A})}{I_2(\sqrt{A})}\left(\frac{U_2}{U_1}-1\right)+\sum_{k=1}^{\infty}\left[\frac{\Omega_k^{(1)}\,\mathrm{sh}\,\beta_k(l-\xi)+\Omega_k^{(2)}\,\mathrm{sh}\,\beta_k\xi}{\mathrm{sh}\,\beta_k\,l}-\right.$$

$$\left.-\frac{\alpha\left[\theta_k^{(1)}\mathrm{sh}\sqrt{A+\beta_k^2}\,(l-\xi)+\theta_k^{(2)}\,\mathrm{sh}\sqrt{A+\beta_k^2}\,\xi\right]}{\mathrm{sh}\sqrt{A+\beta_k^2}\,l}\right]\frac{J_0(\beta_k\eta)-J_0(\beta_k)}{1+\alpha}.$$

The transverse velocities are then expressed as

$$v_1=\sum_{k=1}^{\infty}\left[\frac{\sqrt{A+\beta_k^2}\left(\theta_k^{(1)}\,\mathrm{ch}\sqrt{A+\beta_k^2}\,(l-\xi)-\theta_k^{(2)}\,\mathrm{ch}\sqrt{A+\beta_k^2}\,\xi\right)}{\mathrm{sh}\sqrt{A+\beta_k^2}\,l}+\right.$$

$$\left.+\frac{\beta_k\left(\Omega_k^{(1)}\,\mathrm{ch}\,\beta_k(l-\xi)-\Omega_k^{(2)}\,\mathrm{ch}\,\beta_k\xi\right)}{\mathrm{sh}\,\beta_k\,l}\right]\frac{J_1(\beta_k\eta)-\eta J_1(\beta_k)}{\beta_k(1+\alpha)},$$

$$v_2 = \sum_{k=1}^{\infty} \left[\frac{\beta_k \left(\Omega_k^{(1)} \operatorname{ch} \beta_k (l - \xi) - \Omega_k^{(2)} \operatorname{ch} \beta_k \xi \right)}{\operatorname{sh} \sqrt{A + \beta_k^2}\, l} - \right.$$

$$\left. - \frac{\alpha \sqrt{A + \beta_k^2} \left(\Theta_k^{(1)} \operatorname{ch} \sqrt{A + \beta_k^2} (l - \xi) - \Theta_k^{(2)} \operatorname{ch} \sqrt{A + \beta_k^2}\, \xi \right)}{\operatorname{sh} \sqrt{A + \beta_k^2}\, l} \right] \cdot \frac{J_1 \left(\beta_k \eta \right) - \eta J_1 \left(\beta_k \right)}{\beta_k (1 + \alpha)},$$

and the pressure will be

$$q = q^{\circ} - \frac{8}{1 + \alpha} \left(\alpha + \frac{U_2}{U_1} \right) \xi + 2 \sqrt{A}\, \xi\, \frac{I_1 (\sqrt{A})}{I_2 (\sqrt{A})\, (1 + \alpha)} \left(\frac{U_2}{U_1} - 1 \right) +$$

$$+ 2 \sum_{k=1}^{\infty} \left[\frac{\Theta_k^{(1)} \operatorname{ch} \sqrt{A + \beta_k^2} (l - \xi) - \Theta_k^{(2)} \operatorname{ch} \sqrt{A + \beta_k^2}\, \xi}{\sqrt{A + \beta_k^2} \operatorname{sh} \sqrt{A + \beta_k^2}\, l} + \frac{\Omega_k^{(1)} \operatorname{ch} \beta_k (l - \xi) - \Omega_k^{(2)} \operatorname{ch} \beta_k \xi}{\beta_k \operatorname{sh} \beta_k l} \right] \frac{\beta_k\, J_1 \left(\beta_k \right)}{1 + \alpha} + K' \left(\frac{U_2}{U_1} + 1 \right) \xi.$$

We find the tangential stresses as

$$\tau = - \frac{4}{1 + \alpha} \left(\alpha + \frac{U_2}{U_1} \right) + \frac{\sqrt{A}\, I_1 (\sqrt{A})}{I_2 (\sqrt{A})(1 + \alpha)} \left(\frac{U_2}{U_1} - 1 \right) -$$

$$- \sum_{k=1}^{\infty} \left[\frac{\Theta_k^{(1)} \operatorname{sh} \sqrt{A + \beta_k^2} (l - \xi) + \Theta_k^{(2)} \operatorname{sh} \sqrt{A + \beta_k^2}\, \xi}{\operatorname{sh} \sqrt{A + \beta_k^2}\, l} + \frac{\Omega_k^{(1)} \operatorname{sh} \beta_k (l - \xi) + \Omega_k^{(2)} \operatorname{sh} \beta_k \xi}{\operatorname{sh} \beta_k l} \right] \frac{\beta_k\, J_1 \left(\beta_k \right)}{1 + \alpha},$$

$$\tau_1' = - \frac{4}{1 + \alpha} \left(\alpha + \frac{U_2}{U_1} \right) - \frac{\alpha}{1 + \alpha} \frac{\sqrt{A}\, I_1 (\sqrt{A})}{I_2 (\sqrt{A})} \left(\frac{U_2}{U_1} - 1 \right) -$$

$$- \sum_{k=1}^{\infty} \left[\frac{\Omega_k^{(1)} \operatorname{sh} \beta_k (l - \xi) + \Omega_k^{(2)} \operatorname{sh} \beta_k \xi}{\operatorname{sh} \beta_k l} - \frac{\alpha \left(\Theta_k^{(1)} \operatorname{sh} \sqrt{A + \beta_k^2} (l - \xi) + \Theta_k^{(2)} \operatorname{sh} \sqrt{A + \beta_k^2}\, \xi \right)}{\operatorname{sh} \sqrt{A + \beta_k^2}\, l} \right] \frac{\beta_k\, J_1 \left(\beta_k \right)}{1 + \alpha}.$$

Let us examine the problem of motion in a semi-infinite cylinder. As $l = \infty$ we obtain the following solutions for $\mathrm{Re}_1 = \mathrm{Re}_2$:

$$u_1 = \frac{2}{1 + \alpha} \left(\alpha + \frac{U_2}{U_1} \right) (1 - \eta^2) + \frac{I_0 (\sqrt{A}\, \eta) - I_0 (\sqrt{A})}{I_2 (\sqrt{A})\, (1 + \alpha)} \left(\frac{U_2}{U_1} - 1 \right) +$$

$$+ \sum_{k=1}^{\infty} \left(\Theta_k e^{-n_k \xi} + \Omega_k e^{-\gamma_k \xi} \right) \frac{J_0 \left(\beta_k \eta \right) - J_0 \left(\beta_k \right)}{1 + \alpha},$$

$$u_2 = \frac{2}{1 + \sigma} \left(\alpha + \frac{U_2}{U_1} \right) (1 - \eta^2) - \frac{\alpha}{1 + \alpha} \frac{I_0 (\sqrt{A}\, \eta) - I_0 (\sqrt{A})}{I_2 (\sqrt{A})} \times$$

$$\times \left(\frac{U_2}{U_1} - 1 \right) + \sum_{k=1}^{\infty} \left(\Omega_k e^{-\gamma_k \xi} - \alpha \Theta_k e^{-n_k \xi} \right) \frac{J_0 \left(\beta_k \eta \right) - J_0 \left(\beta_k \right)}{1 + \alpha}.$$

$$v_1 = \sum_{k=1}^{\infty} \left(n_k \Theta_k e^{-n_k \xi} + \gamma_k \Omega_k e^{-\gamma_k \xi} \right) \frac{J_1 \left(\beta_k \eta \right) - \eta J_1 \left(\beta_k \right)}{\beta_k (1 + \alpha)},$$

$$v_2 = \sum_{k=1}^{\infty} \left(\gamma_k \Omega_k e^{-\gamma_k \xi} - \alpha n_k \Theta_k e^{-n_k \xi} \right) \frac{J_1 \left(\beta_k \eta \right) - \eta J_1 \left(\beta_k \right)}{\beta_k (1 + \alpha)},$$

$$q = q^0 - \frac{8}{1+\alpha}\left(\alpha + \frac{U_2}{U_1}\right)\xi + 2\sqrt{A}\xi\,\frac{I_1(\sqrt{A})}{I_2(\sqrt{A})(1+\alpha)}\left(\frac{U_2}{U_1}-1\right) +$$

$$+ 2\sum_{k=1}^{\infty}\left(\frac{\Theta_k}{n_k}e^{-n_k\xi} + \frac{\Omega_k}{\gamma_k}e^{-\gamma_k\xi}\right)\frac{\beta_k I_1(\beta_k)}{1+\alpha} + K'\left(\frac{U_2}{U_1}-1\right)\xi,$$

$$\tau = -\frac{4}{1+\alpha}\left(\alpha + \frac{U_2}{U_1}\right) + \frac{\sqrt{A}I_1(\sqrt{A})}{I_2(\sqrt{A})(1+\alpha)}\left(\frac{U_2}{U_1}-1\right) - \sum_{k=1}^{\infty}\left(\Theta_k e^{-n_k\xi} - \Omega_k e^{-\gamma_k\xi}\right)\frac{\beta_k J_1(\beta_k)}{1+\alpha},$$

$$\tau' = -\frac{4}{1+\alpha}\left(\alpha + \frac{U_2}{U_1}\right) - \frac{\alpha}{1+\alpha}\frac{\sqrt{A}I_1(\sqrt{A})}{I_2(\sqrt{A})}\left(\frac{U_2}{U_1}-1\right) - \sum_{k=1}^{\infty}\left(\Omega_k e^{-\gamma_k\xi} - \alpha\Theta_k e^{-n_k\xi}\right)\frac{\beta_k J_1(\beta_k)}{1+\alpha},$$

where Θ_k and Ω_k are Fourier coefficients of the form

$$\Theta_k = \frac{2}{J_1^2(\beta_k)}\int_0^1 \eta(\varphi - \psi)J_0(\beta_k\eta)\,d\eta + \frac{AJ_0(\beta_k\eta)}{J_1^2(\beta_k)(\beta_k^2+A)}\left(\frac{U_2}{U_1}-1\right),$$

$$\Omega_k = \frac{2}{J_1^2(\beta_k)}\int_0^1 \eta(\alpha\varphi + \psi)J_0(\beta_k\eta)\,d\eta;$$

where $\varphi(\eta) = u_1|_{\xi=0}$, $\psi(\eta) = u_2|_{\xi=0}$. It is easy to note that as $\xi \to \infty$ there is Poiseuille-like flow of both components in the conduit. As is customary [23], let us determine the length of the initial sections l_i ($i = 1, 2$) from the relation

$$\frac{u_{i(\infty, 0)} - u_{i(\xi, 0)}}{u_{i(\infty, 0)}} \approx 0.01.$$

Limiting ourselves to the first member of the series in the velocity formulas, and utilizing the preceding equality to determine the length of the initial sections of each medium, we obtain

$$\Theta_k e^{-n_1 l_1} + \Omega_k e^{-\gamma_1 l_1} = \frac{1}{100\,[J_0(\beta_k)-1]}\left[2\left(\alpha + \frac{U_2}{U_1}\right) + \frac{1 - I_0(\sqrt{A})}{I_2(\sqrt{A})}\left(\frac{U_2}{U_1}-1\right)\right],$$

$$\Omega_k e^{-\gamma_1 l_2} - \alpha\Theta_k e^{-n_1 l_2} = \frac{1}{100\,[J_0(\beta_1)-1]}\left[2\left(\alpha + \frac{U_2}{U_1}\right) - \alpha\frac{1 - I_0(\sqrt{A})}{I_2(\sqrt{A})}\left(\frac{U_2}{U_1}-1\right)\right].$$

Since there is an interaction force between the moving phases, it is possible to put $l_1 \approx l_2 = l$ in the case we consider. We then have

$$l \approx \frac{1}{\gamma_1}\ln\left[50\,|\Omega_1|\,\frac{|J_0(\beta_1)-1|}{\alpha + U_2/U_1}\right].$$

Computations have shown that the initial sections are of identical length in the case of a viscous two-phase flow with $Re_1 \approx Re_2$ (very large) and single-phase incompressible fluid flow.

A study of the dependence of the length of the initial section on the motion parameters for different two-phase flows is of great theoretical and practical interest.

M. P. Nazarii has examined the development of laminarity in the flow of incompressible viscous two-phase media for the case $Re_1 \neq Re_2$. He applied the Laplace–Carson transformation to the system (22.7), then solved the resulting system of ordinary differential equations by the d'Alembert method. By computing the length of the initial section by starting from the formula obtained in such a way, it can be seen that if the fluid discharge in the case of single-phase motion in a cylindrical conduit equals the discharge of this fluid in a two-phase flow along the same conduit, where the second phase is less viscous, then the length of the initial section of the two-phase flow will be longer than that of the single-phase flow.

§23. Combined Two-Phase Motion of a Viscous and an Ideal Medium in Finite and Infinite Plane Conduits

As is known, gases are considered ideal in aerodynamics; many theoretical results and deductions obtained under such an assumption agree with experiment. Therefore, natural gas with water or petroleum particles may also be considered as a visco–ideal two-phase medium, by which term we mean a mixture in which one phase is viscous and the other ideal. Let us now elucidate the development of laminar flow of such media between parallel walls by starting from the approximate equations of motion of visco–ideal two-phase media in a finite plane conduit. On the basis of (6.2) and (6.3) with $\mu_1 = 0$ and $\mu_2 = \mu$, the following system of equations can be obtained:

for the first (ideal) medium

$$\rho_1 \frac{du_1}{dt} = -f_1 \frac{\partial p}{\partial x} + K(u_2 - u_1),$$

$$\rho_1 \frac{dv_1}{\partial t} = -f_1 \frac{\partial p}{\partial y} + K(v_2 - v_1);$$

for the second (viscous) medium

$$\rho_2 \frac{du_2}{dt} = -f_2 \frac{\partial p}{\partial x} + \frac{\partial}{\partial x} f_2 \mu \left(2\frac{\partial u_2}{\partial x} - \frac{2}{3} \operatorname{div} \vec{V}_2 \right) + \frac{\partial}{\partial y} f_2 \mu \left(\frac{\partial u_2}{\partial y} + \frac{\partial v_2}{\partial x} \right) + K(u_1 - u_2),$$

$$\rho_2 \frac{\partial v_2}{\partial t} = -f_2 \frac{\partial p}{\partial y} + \frac{\partial}{\partial y} f_2 \mu \left(2 \frac{\partial v_2}{\partial y} - \frac{2}{3} \operatorname{div} \vec{V}_2 \right) + \frac{\partial}{\partial x} f_2 \mu \left(\frac{\partial u_2}{\partial y} + \frac{\partial v_2}{\partial x} \right) + K(v_1 - v_2).$$

Let us assume that the motion is steady, both media are incompressible, the porosities of the media are constant, and each medium has its own pressure. Then the above-mentioned system of equations becomes

$$\left.
\begin{aligned}
u_1 \frac{\partial u_1}{\partial x} + v_1 \frac{\partial u_1}{\partial y} &= -\frac{1}{\rho_1} \frac{\partial p_1}{\partial x} + \frac{K}{\rho_1}(u_2 - u_1) \\[4pt]
u_1 \frac{\partial v_1}{\partial x} + v_1 \frac{\partial v_1}{\partial y} &= -\frac{1}{\rho_1} \frac{\partial p_1}{\partial y} + \frac{K}{\rho_1}(v_2 - v_1) \\[4pt]
u_2 \frac{\partial u_2}{\partial x} + v_2 \frac{\partial u_2}{\partial y} &= -\frac{1}{\rho_2} \frac{\partial p_2}{\partial x} + \nu \Delta u_2 + \frac{K}{\rho_2}(u_1 - u_2) \\[4pt]
u_2 \frac{\partial v_2}{\partial x} + v_2 \frac{\partial v_2}{\partial y} &= -\frac{1}{\rho_2} \frac{\partial p_2}{\partial y} + \nu \Delta v_2 + \frac{K}{\rho_2}(v_1 - v_2)
\end{aligned}
\right\}
\qquad (23.1)$$

where $\nu = \mu/\rho_{2i}$ is the kinematic viscosity of the second medium.

Let us note that the second member in the continuity equation (3.2) should be discarded in one-dimensional motion of an ideal medium. Since the ideal medium moves in conjunction with the viscous medium in our investigation, and there is an interaction force between them, the second member $\partial v_1/\partial y$ is retained.

Let us apply all the operations (for estimation of the terms) made in § 21 to (23.1). We hence obtain the system of equations

$$\left.\begin{aligned}
u_1 \frac{\partial u_1}{\partial x} + v_1 \frac{\partial u_1}{\partial y} &= -\frac{1}{\rho_1}\frac{\partial p_1}{\partial x} + \frac{K}{\rho_1}(u_2 - u_1) \\[6pt]
\frac{\partial p_1}{\partial y} &= 0, \quad \frac{\partial u_1}{\partial x} + \frac{\partial v_1}{\partial y} = 0 \\[6pt]
u_2 \frac{\partial u_2}{\partial x} + v_2 \frac{\partial u_2}{\partial y} &= -\frac{1}{\rho_2}\frac{\partial p_2}{\partial x} + \nu\Delta u_2 + \frac{K}{\rho_2}(u_1 - u_2) \\[6pt]
\frac{\partial p_2}{\partial y} &= 0, \quad \frac{\partial u_2}{\partial x} + \frac{\partial v_2}{\partial y} = 0
\end{aligned}\right\} \quad (23.2)$$

which is written, after linearization according to Oseen, as

$$\left.\begin{aligned}
U_1 \frac{\partial u_1}{\partial x} &= -\frac{1}{\rho_1}\frac{\partial p_1}{\partial x} + \frac{K}{\rho_1}(u_2 - u_1) \\[6pt]
\frac{\partial p_1}{\partial y} &= 0, \quad \frac{\partial u_1}{\partial x} + \frac{\partial v_2}{\partial y} = 0 \\[6pt]
U_2 \frac{\partial u_2}{\partial x} &= -\frac{1}{\rho_2}\frac{\partial p_2}{\partial x} + \nu\Delta u_2 + \frac{K}{\rho_2}(u_1 - u_2) \\[6pt]
\frac{\partial p_2}{\partial y} &= 0, \quad \frac{\partial u_2}{\partial x} + \frac{\partial v_2}{\partial y} = 0
\end{aligned}\right\} \quad (23.3)$$

The boundary conditions will be

$$\left.\begin{aligned}
\text{for } y = \pm h \qquad & x > 0, \; u_2 = 0, \; v_2 = 0 \\
\text{for } x = 0 \qquad & u_1 = U_1, \; u_2 = U_2 \\
\text{for } x = L \qquad & u_1 = \varphi_1(y), \; u_2 = \varphi_2(y)
\end{aligned}\right\} \quad (23.4)$$

The conditions of constancy of the discharges are expressed thus:

$$\left.\begin{aligned}
\int_{-h}^{h} u_1 \, dy &= 2hU_1 \\[6pt]
\int_{-h}^{h} u_2 \, dy &= 2hU_2
\end{aligned}\right\} \quad (23.5)$$

Let us introduce dimensionless parameters in place of the dimensional quantities in the system (23.3):

for the first medium

$$\left.\begin{aligned}
u_1 &= U_1 u, \; v_2 = U_1 v, \; x = hx_1, \; y = y_1 h \\[6pt]
p_1 &= \rho_1 U_1^2 \, q_1, \; K = \frac{\rho_1 U_1}{h} K_1
\end{aligned}\right\} \quad (23.6)$$

for the second medium

$$\left.\begin{aligned}
u_2 &= U_1 u', \; v_2 = U_1 v' \\[6pt]
p_2 &= \frac{\rho_2 U_1 U_2}{\text{Re}} \, q_2, \; \text{Re} = U_2 h/\nu
\end{aligned}\right\} \quad (23.7)$$

The system (23.3) in dimensionless quantities is written

$$
\left.
\begin{aligned}
\frac{\partial u}{\partial x_1} &= -\frac{\partial q_1}{\partial x_1} + K_1(u' - u) \\[6pt]
\frac{\partial q_1}{\partial y} &= 0 \\[6pt]
\frac{\partial u}{\partial x_1} + \frac{\partial v}{\partial y_1} &= 0 \\[6pt]
\operatorname{Re}\frac{\partial u'}{\partial x_1} &= -\frac{\partial q_2}{\partial x_1} + \Delta u' + K_2(u - u') \\[6pt]
\frac{\partial q_2}{\partial y} &= 0, \; \frac{\partial u'}{\partial x_1} + \frac{\partial v'}{\partial y_1} = 0
\end{aligned}
\right\}
\tag{23.8}
$$

where

$$
K_2 = \frac{\rho_1 U_1}{\rho_2 U_2}\operatorname{Re} K_1.
$$

Then the boundary conditions (23.4) become

$$
\left.
\begin{aligned}
&\text{for } y_1 = \pm 1 && x_1 > 0,\; u' = 0,\; v' = 0 \\
&\text{for } x_1 = 0 && u = 1,\; u' = U_2/U_1 \\
&\text{for } x_1 = L/h = l && u = \psi_1(y_1),\; u' = \psi_2(y_1)
\end{aligned}
\right\}
\tag{23.9}
$$

and the conditions (23.5) for constancy of the discharges are

$$
\left.
\begin{aligned}
\int_0^1 u\, dy_1 &= 1 \\[6pt]
\int_0^1 u'\, dy_1 &= U_2/U_1
\end{aligned}
\right\}
\tag{23.10}
$$

Utilizing the first relationship in (23.10) and the second equation in (23.8), we find

$$
\frac{\partial q_1}{\partial x_1} = K_1(U_2/U_1 - 1).
\tag{23.11}
$$

By using the second relationship in (23.10) we have from the fifth equation in (23.8)

$$
\frac{\partial q_2}{\partial x_1} = \left(\frac{\partial u'}{\partial y_1}\right)_{y_1=1} + K_2\left(1 - \frac{U_2}{U_1}\right).
\tag{23.12}
$$

Comparison of (23.11) and (23.12) confirms our assumption concerning the existence of different pressures for each medium. Otherwise, as is seen from (23.12), the velocity u' would be independent of x_1.

Substituting (23.11) and (23.12) into (23.8), we obtain the system of differential equations

$$
\left.
\begin{aligned}
\frac{\partial u}{\partial x_1} &= K_1(u' - u) - K_1\left(\frac{U_2}{U_1} - 1\right) \\[6pt]
\operatorname{Re}\frac{\partial u'}{\partial x_1} &= \Delta u' + K_2(u - u') - \left(\frac{\partial u'}{\partial y_1}\right)_{y_1=1} - K_2\left(1 - \frac{U_2}{U_1}\right)
\end{aligned}
\right\}
\tag{23.13}
$$

Therefore, the motion of a visco-ideal two-phase medium in a finite plane conduit is described by the system of approximate equations (23.13).

Let us consider the system (23.13) for a semi-infinite conduit; here terms containing the factor δ^2 and higher orders can be discarded, and it then becomes

$$\frac{\partial u}{\partial x_1} = K_1(u'-u) - K_1\left(\frac{U_2}{U_1}-1\right)$$
$$\mathrm{Re}\frac{\partial u'}{\partial x_1} = \frac{\partial^2 u'}{\partial y_1^2} + K_2(u-u') - \left(\frac{\partial u'}{\partial y_1}\right)_1 - K_2\left(1-\frac{U_2}{U_1}\right)$$
(23.14)

The boundary conditions are

for $y_1 = \pm 1$ $\qquad x_1 > 0, \; u' = 0$

for $x_1 = 0$ $\qquad u = 1, \; u' = \frac{U_2}{U_1}$
(23.15)

In order to solve the system (23.14) with the boundary conditions (23.15), let us apply the method of the functional Laplace transform. To do this, we transform from the originals $u(x,y)$ and $u'(x,y)$ to the transforms $\bar{u}(y,\bar{p})$ and $\bar{u}'(y,\bar{p})$ by means of the following substitutions:*

$$u \to \frac{\bar{u}}{\bar{p}}, \quad u' \to \frac{\bar{u}'}{\bar{p}}$$
$$\frac{\partial u}{\partial x} \to -(u)_{x=0} + \bar{u} = -1 + \bar{u}$$
$$\frac{\partial u'}{\partial x} \to -(u')_{x=0} + \bar{u}' - \frac{U_2}{U_1} + \bar{u}'$$
$$\frac{\partial u'}{\partial y} \to \frac{1}{\bar{p}}\frac{d\bar{u}'}{dy}, \quad \frac{\partial^2 u'}{\partial y^2} \to \frac{1}{\bar{p}}\frac{d^2\bar{u}'}{dy^2}$$
(23.16)

Substituting (23.16) into (23.14), we have

$$(\bar{p}+K_1)\bar{u} - K_1\bar{u}' = p - K_1\left(\frac{U_2}{U_1}-1\right)$$
$$\frac{d^2\bar{u}'}{dy^2} - (\bar{p}\mathrm{Re}+K_2)\bar{u}' + K_2\bar{u} = -(\bar{p}\mathrm{Re}+K_2)\frac{U_2}{U_1} + K_2 + \left(\frac{d\bar{u}'}{dy}\right)_{y=1}$$
(23.17)

The boundary conditions for the transforms remain the same as for the originals, i.e.,

for $y = \pm 1$ $\qquad x > 0, \; \bar{u}' = 0$

for $x = 0$ $\qquad \bar{u} = 1, \; \bar{u}' = U_2/U_1$
(23.18)

We find $\bar{u}$ from the first equation in (23.17):

$$\bar{u} = \frac{K_1}{\bar{p}+K_1}\bar{u}' + \frac{\bar{p}}{\bar{p}+K_1} - \frac{K_1}{\bar{p}+K_1}\left(\frac{U_2}{U_1}-1\right).$$
(23.19)

* For brevity, we omit the subscript 1 from x_1 and y_1.

Substituting (23.19) into the second equation in (23.17), we obtain

$$\frac{d^2\bar{u}'}{dy^2} - a\bar{u}' = -a\frac{U_2}{U_1} + \left(\frac{d\bar{u}'}{dy}\right)_{y=1}, \tag{23.20}$$

where

$$a = \frac{\operatorname{Re}\bar{p}^2 + (K_2 + K_1\operatorname{Re})\bar{p}}{\bar{p} + K_1}. \tag{23.21}$$

The solution of the differential equation (23.20) is represented as

$$\bar{u}' = c_1\operatorname{sh}\sqrt{a}y + c_2\operatorname{ch}\sqrt{a}y + \frac{U_2}{U_1} - \frac{1}{a}\left(\frac{d\bar{u}'}{dy}\right)_{y=1}. \tag{23.22}$$

Differentiating both sides of (23.22) with respect to y and substituting y = 1 into the expression obtained, we have

$$\left(\frac{d\bar{u}'}{dy}\right)_{y=1} = \sqrt{a}\,(c_1\operatorname{ch}\sqrt{a} + c_2\operatorname{sh}\sqrt{a}). \tag{23.23}$$

Using the boundary conditions (23.18) and the equality (23.23), we find the values of the arbitrary constants c_1 and c_2:

$$\left.\begin{array}{l} c_1 = 0 \\[2mm] c_2 = -\dfrac{U_2}{U_1}\dfrac{1}{\operatorname{ch}\sqrt{a} - \sqrt{\dfrac{1}{a}}\operatorname{sh}\sqrt{a}} \end{array}\right\} \tag{23.24}$$

Keeping (23.23) and (23.24) in mind, we obtain from (23.22)

$$\bar{u}' = \frac{U_2}{U_1}\frac{\operatorname{ch}\sqrt{a}y - \operatorname{ch}\sqrt{a}}{\sqrt{\dfrac{1}{a}}\operatorname{sh}\sqrt{a} - \operatorname{ch}\sqrt{a}}. \tag{23.25}$$

Substituting (23.25) into (23.19), we determine

$$\bar{u} = \frac{K_1}{\bar{p} + K_1}\left[\frac{U_2}{U_1}\frac{\operatorname{ch}\sqrt{a}y - \operatorname{ch}\sqrt{a}}{\sqrt{\dfrac{1}{a}}\operatorname{sh}\sqrt{a} - \operatorname{ch}\sqrt{a}} + c\right], \tag{23.26}$$

where

$$c = \frac{\bar{p}}{K_1} - \left(\frac{U_2}{U_1} - 1\right). \tag{23.27}$$

With the aid of known formulas of operational calculus [8], we now transform from the transforms (23.25) and (23.26) to the originals:

$$u'(x,\,y) = \frac{1}{2\pi i}\frac{U_2}{U_1}\int\limits_{\sigma-i\infty}^{\sigma+i\infty} e^{\bar{p}x}\frac{\operatorname{ch}\sqrt{a}y - \operatorname{ch}\sqrt{a}}{\sqrt{\dfrac{1}{a}}\operatorname{sh}\sqrt{a}\cdot\operatorname{ch}\sqrt{a}}\frac{d\bar{p}}{\bar{p}}, \tag{23.28}$$

$$u(x, y) = \frac{1}{2\pi i} \int\limits_{\sigma - i\infty}^{\sigma + i\infty} e^{\bar{p}x} \frac{K_1}{\bar{p} + K_1} \left(\frac{U_2}{U_1} \frac{\mathrm{ch}\,\sqrt{a}\,y - \mathrm{ch}\,\sqrt{a}}{\sqrt{\frac{1}{a}}\,\mathrm{sh}\,\sqrt{a} - \mathrm{ch}\,\sqrt{a}} + c \right) \frac{d\bar{p}}{\bar{p}}. \tag{23.29}$$

Expanding each member in the numerator and denominator of the integrands in (23.28) and (23.29) in series, and retaining only members of second order or lower in the argument, we write

$$\left. \begin{aligned} (\bar{u}')_{\bar{p} \to 0} &= \frac{3}{2} \frac{U_2}{U_1} (1 - y^2) \\ (\bar{u})_{\bar{p} \to 0} &= \frac{3}{2} \frac{U_2}{U_1} (1 - y^2) - \left(\frac{U_2}{U_1} - 1 \right) \end{aligned} \right\} \tag{23.30}$$

Therefore, the profiles of the fundamental velocity distributions infinitely far from the entrance to a plane conduit will be parabolic.

The integrand in (23.28) has a singularity at the point $\bar{p} = 0$, and at points coincident with the roots of the equation

$$\mathrm{th}\,\sqrt{a} = \sqrt{a}. \tag{23.31}$$

The roots of (23.31) are represented as

$$a = \beta_m^2, \tag{23.32}$$

where β_m are roots of the equation $\tan x = x$.

The integrand of (23.29) is decomposed into proper fractions:

$$\frac{F_1(\bar{p})}{F_2(\bar{p})} = \frac{c_0}{\bar{p}} + \sum_{m=1}^{\infty} \frac{c_m}{\bar{p} - \bar{p}_m};$$

where

$$c_0 = \left[\frac{\mathrm{ch}\,\sqrt{a}\,y - \mathrm{ch}\,\sqrt{a}}{\sqrt{\frac{1}{a}}\,\mathrm{sh}\,\sqrt{a} - \mathrm{ch}\,\sqrt{a}} \right]_{\bar{p} \to 0} = \frac{3}{2} (1 - y^2). \tag{23.33}$$

In order to determine the c_m, it is necessary to find the roots of (23.32), which equal

$$\bar{p}_{1m2m} = \frac{-\left(K_2 + K_1\,\mathrm{Re} + \beta_m^2 \right) \pm \sqrt{\left(K_2 + K_1\,\mathrm{Re} + \beta_m^2 \right)^2 - 4K_1\beta_m^2\,\mathrm{Re}}}{2\mathrm{Re}}; \tag{23.34}$$

here the radical is positive (on the basis of the inequality $a^2 + b^2 \geq 2ab$); hence $\bar{p}_{1m}$ and $\bar{p}_{2m}$ are negative real numbers. Denoting the values of $\bar{p}_{1m}$ and $\bar{p}_{2m}$ by $-\gamma_{1m}$ and $-\gamma_{2m}$, we find

$$\left. \begin{aligned} c_{1m} &= \frac{\cos \beta_m y - \cos \beta_m}{f_{1m} \sin \beta_m} \\ c_{2m} &= \frac{\cos \beta_m y - \cos \beta_m}{f_{2m} \sin \beta_m} \end{aligned} \right\} \tag{23.35}$$

where

$$f_{1m} = -\frac{\gamma_{1m}}{\beta_m} \frac{\operatorname{Re}(\gamma_{1m} + K_1)^2 + K_1 K_2}{(\gamma_{1m} + K_1)^2},$$

$$f_{2m} = -\frac{\gamma_{2m}}{\beta_m} \frac{\operatorname{Re}(\gamma_{2m} + K_1)^2 + K_1 K_2}{(\gamma_{2m} + K_1)^2}.$$

Taking account of (23.33) and (23.35), we finally have

$$u'(x,\ y) = \frac{3}{2} \frac{U_2}{U_1} (1 - y^2) + 2\frac{U_2}{U_1} \sum_{m=1}^{\infty} \left(\frac{e^{\gamma_{1m} x}}{f_{1m}} + \frac{e^{\gamma_{2m} x}}{f_{2m}} \right) \frac{\cos \beta_m y - \cos \beta_m}{\sin \beta_m}. \qquad (23.36)$$

Proceeding analogously with the integrand in (23.29), we find

$$u(x,\ y) = \frac{3}{2} \frac{U_2}{U_1} (1 - y^2) - \left(\frac{U_2}{U_1} - 1 \right) - \left[\frac{3}{2} \frac{U_2}{U_1} (1 - y^2) - \frac{U_2}{U_1} \right] e^{-K_1 x} +$$

$$+ 2\frac{U_2}{U_1} \sum_{m=1}^{\infty} \left(\frac{e^{\gamma_{1m} x}}{f'_{1m}} + \frac{e^{\gamma_{2m} x}}{f'_{2m}} \right) \frac{\cos \beta_m y - \cos \beta_m}{\sin \beta_m}, \qquad (23.37)$$

where

$$\left. \begin{array}{l} f'_{1m} = \dfrac{\gamma_{1m} + K_1}{K_1} f_{1m} \\[2mm] f'_{2m} = \dfrac{\gamma_{2m} + K_1}{K_1} f_{2m} \end{array} \right\}. \qquad (23.38)$$

From (23.37) as a particular case for an incompressible viscous fluid [10] we obtain for K = 0

$$u(x,\ y) = \frac{3}{2} (1 - y^2) + 2 \sum_{m=1}^{\infty} \frac{\cos \beta_m y - \cos \beta_m}{\beta_m^2 \cos \beta_m} e^{-\frac{\beta_m^2}{\operatorname{Re}} x}.$$

Now, utilizing (23.36) and (23.37), we may find expressions for the other motion parameters. We find v and v' from the continuity equation:

$$\left. \begin{array}{l} v(x,\ y) = \dfrac{K_1}{2} \dfrac{U_2}{U_1} (y^2 - 1) y e^{-K_1 x} + 2\dfrac{U_2}{U_1} \sum_{m=1}^{\infty} \left(\dfrac{\gamma_{1m}}{f'_{1m}} e^{\gamma_{1m} x} + \dfrac{\gamma_{2m}}{f'_{2m}} e^{\gamma_{2m} x} \right) \dfrac{y \sin \beta_m - \sin \beta_m y}{\beta_m \sin \beta_m} \\[4mm] v'(x,\ y) = 2\dfrac{U_2}{U_1} \sum_{m=1}^{\infty} \left(\dfrac{\gamma_{1m}}{f_{1m}} e^{\gamma_{1m} x} + \dfrac{\gamma_{2m}}{f_{2m}} e^{\gamma_{2m} x} \right) \dfrac{y \sin \beta_m - \sin \beta_m y}{\beta_m \sin \beta_m} \end{array} \right\} \qquad (23.39)$$

We find the pressure from the relationships (23.11) and (23.12):

$$\left. \begin{array}{l} q_1(x) = q_0 + K_1 (U_2/U_1 - 1) x \\[3mm] q_2(x) = q'_0 - \dfrac{3}{2} \dfrac{U_2}{U_1} x + K_2 \left(1 - \dfrac{U_2}{U_1} \right) x - 2\dfrac{U_2}{U_1} \sum_{m=1}^{\infty} \beta_m \left(\dfrac{e^{\gamma_{1m} x}}{\gamma_{1m} f_{1m}} + \dfrac{e^{\gamma_{2m} x}}{f_{2m} \gamma_{2m}} \right) \end{array} \right\}. \qquad (23.40)$$

Let us determine the dimensionless lengths of the initial sections l_1 and l_2. To this end, let us form an expression for each medium:

$$\frac{u(\infty,\,0)-u(l_1,\,0)}{u(\infty,\,0)} = \left[\frac{1}{2}\cdot e^{-K_1 l} + 2\sum_{m=1}^{\infty}\left(\frac{e^{\gamma_{1m}l_1}}{f'_{1m}} + \frac{e^{\gamma_{2m}l_1}}{f'_{2m}}\right)\frac{\cos\beta_m - 1}{\sin\beta_m}\right]\frac{1}{\frac{1}{2}+\frac{U_1}{U_2}}, \tag{23.41}$$

$$\frac{u'(\infty,\,0)-u'(l_2,\,0)}{u'(\infty,\,0)} = \frac{4}{3}\sum_{m=1}^{\infty}\left(\frac{e^{\gamma_{1m}l_2}}{f_{1m}} + \frac{e^{\gamma_{2m}l_2}}{f_{2m}}\right)\frac{\cos\beta_m - 1}{\sin\beta_m}. \tag{23.42}$$

Assigning values to the left sides of (23.41) and (23.42), and solving the resulting equations for l_1 and l_2, approximate values may be found for the lengths of the initial sections of a plane conduit in the case under consideration. Let the left side of (23.41) and (23.42) equal the small number α. Then

$$\left.\begin{array}{c} 2\sum_{m=1}^{\infty}\left(\dfrac{e^{\gamma_{1m}l_1}}{f'_{1m}} + \dfrac{e^{\gamma_{2m}l_1}}{f'_{2m}}\right)\dfrac{\cos\beta_m - 1}{\sin\beta_m} + \dfrac{1}{2}\cdot e^{-K_1 l_1} = \alpha\left(\dfrac{1}{2}+\dfrac{U_1}{U_2}\right) \\[3ex] \dfrac{4}{3}\sum_{m=1}^{\infty}\left(\dfrac{e^{\gamma_{1m}l_2}}{f_{1m}} + \dfrac{e^{\gamma_{2m}l_2}}{f_{2m}}\right)\dfrac{\cos\beta_m - 1}{\sin\beta_m} = \alpha \end{array}\right\} \tag{23.43}$$

For a single-phase flow we have the following known formula from the second equation in (23.43)

$$e^{-\frac{\beta_1^2}{h Re_0}} = \frac{3}{4}\,\alpha\,\frac{\beta_1^2\cos\beta_1}{\cos\beta_1 - 1}; \tag{23.44}$$

where $Re_0 = Uh/\nu$ is the Reynolds number of single-phase motion.

If an expression is found for L_0 from (23.44), then for an incompressible viscous fluid with $\alpha = 0.01$

$$L_0 = 0.18\ h Re_0,$$

which agrees with the results in [10].

We find approximate values for the lengths of the initial sections from (23.43). Retaining only the first members of the sum in the left sides of (23.43), we have

$$\left.\begin{array}{c} \dfrac{1}{f'_{11}}e^{\gamma_{11}l_1} + \dfrac{1}{f'_{21}}e^{\gamma_{21}l_1} + \dfrac{1}{4}\dfrac{\beta_1\cos\beta_1}{\cos\beta_1-1}\cdot e^{-K_1 l_1} = \dfrac{1}{2}\,\alpha\,\dfrac{\beta_1\cos\beta_1}{\cos\beta_1-1}\left(\dfrac{1}{2}+\dfrac{U_1}{U_2}\right) \\[3ex] \dfrac{1}{f_{11}}e^{\gamma_{11}l_2} + \dfrac{1}{f_{21}}e^{\gamma_{21}l_2} = \dfrac{3}{4}\,\alpha\,\dfrac{\beta_1\cos\beta_1}{\cos\beta_1-1} \end{array}\right\} \tag{23.45}$$

Calculations show that all the members, except the first, are small in the left sides of (23.45) for any values of the parameters $K > 0, f_1, f_2$; hence to evaluate the approximate values of the dimensional lengths of the initial sections of two-phase motion of a visco–ideal medium, we obtain the following formulas from (23.45):

$$L_1 = \frac{h}{|\gamma_{11}|}\ln\frac{4}{f'_{11}\alpha}\frac{U_2}{U_2 - 2U_1}\frac{\cos\beta_1 - 1}{\beta_1\cos\beta_1}, \tag{23.46}$$

for the ideal phase, and

$$L_2 = \frac{h}{|\gamma_{11}|} \ln \frac{4}{3f_{11}\pi} \frac{\cos \beta_1 - 1}{\beta_1 \cos \beta_1} \tag{23.47}$$

for the viscous phase.

Calculations have shown that (23.47) may be written as

$$L_2 \approx 0.18\, h \mathrm{Re}. \tag{23.48}$$

Moreover, we obtain

$$U_2 = U/f_2$$

for the identical discharges of single-phase motion of a viscous incompressible fluid and the same fluid in a visco–ideal two-phase flow, where U is the mean discharge velocity of the single-phase motion of the viscous medium. Then (23.48) is written as

$$L_2 = 0.18h\frac{\mathrm{Re}_0}{f_2}, \quad \text{or} \quad L_0/L_2 = f_2. \tag{23.49}$$

Therefore, the length of the initial section for the viscous phase of a visco–ideal flow is greater than the length of the initial section for the same fluid in single-phase flow.

§ 24. Motion of Incompressible Viscous Three-Phase Media in a Plane Conduit

Let us consider the motion of incompressible viscous three-phase media between two fixed parallel walls extending to infinity on both sides. We consider 2h to be the spacing between the walls, and the 0x axis to be directed along the axis of the plane conduit [24].

Let the motion be rectilinear and steady and the porosities constant. Then the equations of motion for three-phase viscous media and the equations of continuity (5.1), (5.2), (5.3), and (3.2) become

$$\left.\begin{aligned}
f_1\mu_1\frac{d^2u}{dy^2} + K_{21}(u_2 - u_1) + K_{31}(u_3 - u_1) &= f_1 N \\[4pt]
f_2\mu_2\frac{d^2u_2}{dy^2} + K_{12}(u_1 - u_2) + K_{32}(u_3 - u_2) &= f_2 N \\[4pt]
f_3\mu_3\frac{d^2u_3}{dy^2} + K_{31}(u_1 - u_3) + K_{23}(u_2 - u_3) &= f_3 N \\[4pt]
\frac{\partial u_1}{\partial x} &= 0 \\[4pt]
\frac{\partial u_2}{\partial x} &= 0 \\[4pt]
\frac{\partial u_3}{\partial x} &= 0
\end{aligned}\right\} \tag{24.1}$$

The boundary conditions of the problem under consideration will be

$$y = \pm h, \ u_1 = u_2 = u_3 = 0. \tag{24.2}$$

Term-by-term addition of the equations in (24.1) and taking account of the expression (2.1) yield

$$f_{1\mu_1}\frac{d^2u_1}{dy^2} + f_{2\mu_2}\frac{d^2u_2}{dy^2} + f_{3\mu_3}\frac{d^2u_3}{dy^2} = N.$$

Solving this latter equation with the boundary conditions (24.2), we obtain

$$f_{1\mu_1}u_1 + f_{2\mu_2}u_2 + f_{3\mu_3}u_3 = \frac{N}{2}(y^2 - h^2),\tag{24.3}$$

from which

$$u_3 = -\frac{f_{1\mu_1}}{f_{3\mu_3}}u_1 - \frac{f_{2\mu_2}}{f_{3\mu_3}}u_2 + \frac{N}{2f_{2\mu_2}}(y^2 - h^2).\tag{24.4}$$

Adding the second and third, and the first and third equations of (24.1) term by term, and taking account of (24.4), we obtain

$$\left.\begin{aligned}\frac{d^2u_1}{dy^2} - au_1 + bu_2 &= N\left[\frac{1}{\mu_1} - \frac{K_{13}}{2f_{1\mu_1}f_{3\mu_3}}(y^2 - h^2)\right]\\[2mm]\frac{d^2u_2}{dy^2} - cu_2 + du_1 &= N\left[\frac{1}{\mu_2} - \frac{K_{23}}{2f_{2\mu_2}f_{3\mu_3}}(y^2 - h^2)\right]\end{aligned}\right\},\tag{24.5}$$

where

$$\left.\begin{aligned}a &= \frac{K_{12} + K_{13}}{f_{1\mu_1}} + \frac{K_{13}}{f_{3\mu_3}}, & b &= \frac{K_{12}}{f_{1\mu_1}} - K_{13}\frac{f_{2\mu_2}}{f_{1\mu_1}f_{3\mu_3}}\\[2mm]c &= \frac{K_{12} + K_{23}}{f_{2\mu_2}} + \frac{K_{23}}{f_{3\mu_3}}, & d &= \frac{K_{12}}{f_{2\mu_2}} - K_{23}\frac{f_{1\mu_1}}{f_{2\mu_2}f_{3\mu_3}}\end{aligned}\right\}.\tag{24.6}$$

We solve the system (24.5) by the d'Alembert method. Multiplying the first equation of this system by some number A and adding the result to the second equation, we obtain a differential equation of the form

$$\frac{d^2}{dy^2}(Au_1 + u_2) + (Ab - c)\left(\frac{-Aa + d}{Ab - c}u_1 + u_2\right) =$$

$$= N\left\{A\left[\frac{1}{\mu_1} - \frac{K_{13}}{2f_{1\mu_1}f_{3\mu_3}}(y^2 - h^2)\right] + \left[\frac{1}{\mu_2} - \frac{K_{13}}{2f_{2\mu_2}f_{3\mu_3}}(y^2 - h^2)\right]\right\}.\tag{24.7}$$

We select the A in (24.7) so that the equality

$$\frac{d - Aa}{Ab - c} = A\tag{24.8}$$

holds. Then (24.7) becomes

$$\frac{d^2}{dy^2}(Au_1 + u_2) - (c - Ab)(Au_1 + u_2) = N\left\{A\left[\frac{1}{\mu_1} - \frac{K_{13}}{2f_{1\mu_1}f_{3\mu_3}}(y^2 - h^2)\right] + \left[\frac{1}{\mu_2} - \frac{K_{23}}{2f_{2\mu_2}f_{3\mu_3}}(y^2 - h^2)\right]\right\}.$$

$$\tag{24.9}$$

The solution of this equation will be

$$Au_1 + u_2 = c_1\,\text{sh}\,\sqrt{c - Ab}\,y + c_2\,\text{ch}\,\sqrt{c - Ab}\,y -$$

$$- \frac{N}{Ab - c}\left\{A\left[\frac{1}{\mu_1} - \frac{K_{13}}{2f_{1\mu_1}f_{3\mu_3}}(y^2 - h^2)\right] + \left[\frac{1}{\mu_2} - \frac{K_{23}}{2f_{2\mu_2}f_{3\mu_3}}(y^2 - h^2)\right]\right\}.\tag{24.10}$$

It follows from the condition (24.8) that there are two values for A:

$$A_{1,2} = \frac{-(a - c) \pm \sqrt{(a - c)^2 + 4bd}}{2b}\tag{24.11}$$

hence, the solution of (24.10) with boundary conditions (24.2) and corresponding values A_1, A_2 will be

$$A_1 u_1 + u_2 = \frac{N}{c - A_1 b} \left\{ \left[\left(\frac{A_1}{\mu_1} - \frac{1}{\mu_2} \right) + \frac{1}{c - A_1 b} \left(\frac{A_1 K_{13}}{f_1 \mu_1} + \frac{K_{23}}{f_2 \mu_2} \right) \frac{1}{f_3 \mu_3} \right] \frac{\operatorname{ch} \sqrt{c - A_1 b}\, y}{\operatorname{ch} \sqrt{c - A_1 b}\, h} - \right.$$
$$- A_1 \left[\frac{1}{\mu_1} - \frac{K_{13}}{2 f_1 \mu_1 f_3 \mu_3} (y^2 - h^2) \right] - \left[\frac{1}{\mu_2} - \frac{K_{23}}{2 f_2 \mu_2 f_3 \mu_3} (y^2 - h^2) \right] + \frac{1}{c - A_1 b} \left[\frac{A_1 K_{13}}{f_1 \mu_1} + \frac{K_{23}}{f_2 \mu_2} \right] \frac{1}{f_3 \mu_3} \right\}, \quad (24.12)$$

$$A_2 u_1 + u_2 = \frac{N}{c - A_2 b} \left\{ \left[\left(\frac{A_2}{\mu_1} - \frac{1}{\mu_2} \right) + \frac{1}{c - A_2 b} \left(\frac{A_2 K_{13}}{f_1 \mu_1} + \frac{K_{23}}{f_2 \mu_2} \right) \frac{1}{f_3 \mu_3} \right] \frac{\operatorname{ch} \sqrt{c - A_2 b}\, y}{\operatorname{ch} \sqrt{c - A_2 b}\, h} - \right.$$
$$- A_2 \left[\frac{1}{\mu_1} - \frac{K_{13}}{2 f_1 \mu_1 f_3 \mu_3} (y^2 - h^2) \right] - \left[\frac{1}{\mu_2} - \frac{K_{23}}{2 f_2 \mu_2 f_3 \mu_3} (y^2 - h^2) \right] + \frac{1}{c - A_2 b} \left[\frac{A_2 K_{13}}{f_1 \mu_1} + \frac{K_{23}}{f_2 \mu_2} \right] \frac{1}{f_3 \mu_3} \right\}. \quad (24.13)$$

We obtain the following formulas for the velocities from (24.12) and (24.13):

$$u_1 = - \frac{N}{A_2 - A_1} \left[\frac{1}{c - A_1 b} \left\{ \left[\left(\frac{A_1}{\mu_1} - \frac{1}{\mu_2} \right) + \frac{1}{c - A_1 b} \left(\frac{A_1 K_{13}}{f_1 \mu_1} + \frac{K_{23}}{f_2 \mu_2} \right) \frac{1}{f_3 \mu_3} \right] \frac{\operatorname{ch} \sqrt{c - A_1 b}\, y}{\operatorname{ch} \sqrt{c - A_1 b}\, h} - A_1 \left[\frac{1}{\mu_1} - \right. \right. \right.$$
$$- \frac{K_{13}}{2 f_1 \mu_1 f_2 \mu_2} (y^2 - h^2) \right] - \left[\frac{1}{\mu_2} - \frac{K_{23}}{2 f_2 \mu_2 f_3 \mu_3} (y^2 - h^2) \right] + \frac{1}{c - A_1 b} \left[\frac{A_1 K_{13}}{f_1 \mu_1} + \frac{K_{23}}{f_2 \mu_2} \right] \frac{1}{f_3 \mu_3} \right\} -$$
$$- \frac{1}{c - A_2 b} \left\{ \left[\left(\frac{A_2}{\mu_1} - \frac{1}{\mu_2} \right) + \frac{1}{c - A_2 b} \left(\frac{A_2 K_{13}}{f_1 \mu_1} + \frac{K_{23}}{f_2 \mu_2} \right) \frac{1}{f_3 \mu_3} \right] \frac{\operatorname{ch} \sqrt{c - A_2 b}\, y}{\operatorname{ch} \sqrt{c - A_2 b}\, h} - A_2 \left[\frac{1}{\mu_1} - \right. \right.$$
$$- \frac{K_{13}}{2 f_1 \mu_1 f_3 \mu_3} (y^2 - h^2) \right] - \left[\frac{1}{\mu_1} - \frac{K_{23}}{2 f_2 \mu_2 f_3 \mu_3} (y^2 - h^2) \right] + \frac{1}{c - A_2 b} \left(\frac{A_2 K_{13}}{f_1 \mu_1} + \frac{K_{23}}{f_2 \mu_2} \right) \frac{1}{f_3 \mu_3} \right\} \right], \quad (24.14)$$

$$u_2 = - \frac{N}{A_1 - A_2} \left[\frac{A_2}{c - A_1 b} \left\{ \left[\left(\frac{A_1}{\mu_1} - \frac{1}{\mu_2} \right) + \frac{1}{c - A_1 b} \left(\frac{A_1 K_{13}}{f_1 \mu_1} + \frac{K_{23}}{f_2 \mu_2} \right) \frac{1}{f_3 \mu_3} \right] \frac{\operatorname{ch} \sqrt{c - A_1 b}\, y}{\operatorname{ch} \sqrt{c - A_1 b}\, h} - \right. \right.$$
$$- A_1 \left[\frac{1}{\mu_1} - \frac{K_{13}}{2 f_1 \mu_1 f_3 \mu_3} (y^2 - h^2) \right] - \left[\frac{1}{\mu_2} - \frac{K_{23}}{2 f_2 \mu_2 f_3 \mu_3} (y^2 - h^2) \right] + \frac{1}{c - A_1 b} \left[\frac{A_1 K_{13}}{f_1 \mu_1} + \right.$$
$$+ \left. \frac{K_{23}}{f_2 \mu_2} \right] \frac{1}{f_3 \mu_3} \right\} - \frac{A_1}{c - A_2 b} \left\{ \left[\left(\frac{A_2}{\mu_1} - \frac{1}{\mu_2} \right) + \frac{1}{c - A_2 b} \left(\frac{A_2 K_{13}}{f_1 \mu_1} + \frac{K_{23}}{f_2 \mu_2} \right) \frac{1}{f_3 \mu_3} \right] \frac{\operatorname{ch} \sqrt{c - A_2 b}\, y}{\operatorname{ch} \sqrt{c - A_2 b}\, h} - \right.$$
$$- A_2 \left[\frac{1}{\mu_1} - \frac{K_{13}}{2 f_1 \mu_1 f_2 \mu_2} (y^2 - h^2) \right] - \left[\frac{1}{\mu_2} - \frac{K_{23}}{2 f_2 \mu_2 f_3 \mu_3} (y^2 - h^2) \right] + \frac{1}{c - A_2 b} \left(\frac{A_2 K_{13}}{f_1 \mu_1} - \frac{K_{23}}{f_2 \mu_2} \right) \frac{1}{f_3 \mu_3} \right\} \right]. \quad (24.15)$$

Then from (24.4) we will have

$$u_3 = - \frac{N}{f_3 \mu_3} \frac{1}{(A_1 - A_2)} \left[\frac{f_1 \mu_1 + A_2 f_2 \mu_2}{c - A_1 b} \left\{ \left[\left(\frac{A_1}{\mu_1} - \frac{1}{\mu_2} \right) + \frac{1}{c - A_1 b} \left(\frac{A_1 K_{13}}{f_1 \mu_1} + \frac{K_{23}}{f_2 \mu_2} \right) \frac{1}{f_3 \mu_3} \right] \times \right. \right.$$
$$\times \frac{\operatorname{ch} \sqrt{c - A_1 b}\, y}{\operatorname{ch} \sqrt{c - A_1 b}\, h} - A_1 \left[\frac{1}{\mu_1} - \frac{K_{13}}{2 f_1 \mu_1 f_3 \mu_3} (y^2 - h^2) \right] - \left[\frac{1}{\mu_2} - \frac{K_{23}}{2 f_2 \mu_2 f_3 \mu_3} (y^2 - h^2) \right] + \frac{1}{c - A_1 b} \times$$
$$\times \left[\frac{A_1 K_{13}}{f_1 \mu_1} + \frac{K_{23}}{f_2 \mu_2} \right] \frac{1}{f_3 \mu_3} \right\} - \frac{f_1 \mu_1 + A_1 f_2 \mu_2}{c - A_2 b} \left\{ \left[\left(\frac{A_2}{\mu_1} - \frac{1}{\mu_2} \right) + \frac{1}{c - A_2 b} \left(\frac{A_2 K_{13}}{f_1 \mu_1} + \frac{K_{23}}{f_2 \mu_2} \right) \frac{1}{f_3 \mu_3} \right] \frac{\operatorname{ch} \sqrt{c - A_2 b}\, y}{\operatorname{ch} \sqrt{c - A_2 b}\, h} - \right.$$
$$- A_2 \left[\frac{1}{\mu_1} - \frac{K_{13}}{2 f_1 \mu_1 f_3 \mu_3} (y^2 - h^2) \right] - \left[\frac{1}{\mu_2} - \frac{K_{23}}{2 f_2 \mu_2 f_3 \mu_3} (y^2 - h^2) \right] + \frac{1}{c - A_2 b} \left(\frac{A_2 K_{23}}{f_1 \mu_1} + \frac{K_{23}}{f_2 \mu_2} \right) \frac{1}{f_3 \mu_3} \right\} \right] \quad (24.16)$$

Let us calculate the volume discharge of each phase of the mixture flowing through the cross section of a plane conduit by means of the formulas

$$Q_1 = f_1 \int_{-h}^{h} u_1 dy, \quad Q_2 = f_2 \int_{-h}^{h} u_2 dy,$$

$$Q_3 = f_3 \int_{-h}^{h} u_3 dy.$$

Substituting the values of u_1, u_2, u_3 in the indicated discharge formulas and integrating, we obtain

$$Q_1 = -\frac{2f_1 N}{A_2 - A_1}\left[\frac{1}{c - A_1 b}\left\{\frac{1}{\sqrt{c - A_1 b}}\left[\left(\frac{A_1}{\mu_1} - \frac{1}{\mu_2}\right) + \frac{1}{c - A_1 b}\left(\frac{A_1 K_{13}}{f_1 \mu_1} + \frac{K_{23}}{f_2 \mu_2}\right)\frac{1}{f_3 \mu_3}\right]\text{th}\,\sqrt{c - A_1 b}\,h - \right.$$

$$\left. - A_1\left[\frac{h}{\mu_1} + \frac{1}{3}\frac{K_{13}h^3}{f_1 \mu_1 f_3 \mu_3}\right] - \left[\frac{h}{\mu_2} + \frac{1}{3}\frac{K_{23}h^3}{f_2 \mu_2 f_3 \mu_3}\right] + \frac{h}{c - A_1 b}\left[\frac{A_1 K_{13}}{f_1 \mu_1} + \frac{K_{23}}{f_2 \mu_2}\right]\frac{1}{f_3 \mu_3}\right\} - $$

$$- \frac{1}{c - A_2 b}\left\{\frac{1}{\sqrt{c - A_2 b}}\left[\left(\frac{A_2}{\mu_1} - \frac{1}{\mu_2}\right) + \frac{1}{c - A_2 b}\left(\frac{A_2 K_{13}}{f_1 \mu_1} + \frac{K_{23}}{f_2 \mu_2}\right)\frac{1}{f_3 \mu_3}\right]\text{th}\,\sqrt{c - A_2 b}\,h - $$

$$\left. - A_2\left[\frac{h}{\mu_1} + \frac{1}{3}\frac{K_{13}h^3}{f_1 \mu_1 f_3 \mu_3}\right] - \left[\frac{h}{\mu_2} + \frac{1}{3}\frac{K_{23}h^3}{f_2 \mu_2 f_3 \mu_3}\right] + \frac{h}{c - A_2 b}\left[\frac{A_2 K_{13}}{f_1 \mu_1} + \frac{K_{23}}{f_2 \mu_2}\right]\frac{1}{f_3 \mu_3}\right\}\right], \quad (24.17)$$

$$Q_2 = -\frac{2f_2 N}{A_1 - A_2}\left[\frac{A_2}{c - A_1 b}\left[\left(\frac{A_1}{\mu_1} - \frac{1}{\mu_2}\right) + \frac{1}{c - A_1 b}\left(\frac{A_1 K_{13}}{f_1 \mu_1} + \frac{K_{23}}{f_2 \mu_2}\right)\frac{1}{f_3 \mu_3}\right] \times$$

$$\times \text{th}\,\sqrt{c - A_1 b}\,h - A_1\left[\frac{h}{\mu_1} + \frac{1}{3}\frac{K_{13}h^3}{f_1 \mu_1 f_3 \mu_3}\right] - \left[\frac{h}{\mu_2} + \frac{1}{3}\frac{K_{23}h^3}{f_2 \mu_2 f_3 \mu_3}\right] + \frac{1}{c - A_1 b}\left[\frac{A_1 K_{13}}{f_1 \mu_1} + \right.$$

$$\left. + \frac{K_{23}}{f_2 \mu_2}\right]\frac{1}{f_3 \mu_3} - \frac{A_1}{c - A_2 b}\left\{\frac{1}{\sqrt{c - A_2 b}}\left[\left(\frac{A_2}{\mu_1} - \frac{1}{\mu_2}\right) + \frac{1}{c - A_2 b}\cdot\left(\frac{A_2 K_{13}}{f_1 \mu_1} + \frac{K_{23}}{f_2 \mu_2}\right)\frac{1}{f_3 \mu_3}\right] \times$$

$$\times \text{th}\,\sqrt{c - A_2 b}\,h - A_2\left[\frac{h}{\mu_1} + \frac{1}{3}\frac{K_{13}h^3}{f_1 \mu_1 f_3 \mu_3}\right] - \left[\frac{h}{\mu_1} + \frac{1}{3}\frac{K_{23}h^3}{f_2 \mu_2 f_3 \mu_3}\right] + \frac{h}{c - A_2 b}\left(\frac{A_2 K_{13}}{f_1 \mu_1} + \frac{K_{23}}{f_2 \mu_2}\right)\frac{1}{f_3 \mu_3}\right\}\right], \quad (24.18)$$

$$Q_3 = \frac{N}{(A_1 - A_2)\mu_3}\left[\frac{f_1 \mu_1 + A_2 f_2 \mu_2}{c - A_1 b}\left\{\frac{1}{\sqrt{c - A_1 b}}\left[\left(\frac{A_1}{\mu_1} - \frac{1}{\mu_2}\right) + \frac{1}{c - A_1 b}\left(\frac{A_1 K_{13}}{f_1 \mu_1} + \frac{K_{23}}{f_2 \mu_2}\right) \times\right.\right.$$

$$\left.\times \frac{1}{f_3 \mu_3}\right]\text{th}\,\sqrt{c - A_1 b}\,h - A_1\left[\frac{h}{\mu_1} + \frac{1}{3}\frac{K_{13}h^3}{f_1 \mu_1 f_3 \mu_3}\right] - \left[\frac{h}{\mu_2} + \frac{1}{3}\frac{K_{23}h^3}{f_2 \mu_2 f_3 \mu_3}\right] + \frac{h}{c - A_1 b}\left[\frac{A_1 K_{13}}{f_1 \mu_1} + \frac{K_{23}}{f_2 \mu_2}\right]\frac{1}{f_3 \mu_3}\right\} - $$

$$- \frac{f_1 \mu_1 + A_1 f_2 \mu_2}{c - A_2 b}\left\{\frac{1}{\sqrt{c - A_2 b}}\left[\left(\frac{A_2}{\mu_1} - \frac{1}{\mu_2}\right) + \frac{1}{c - A_2 b}\left(\frac{A_2 K_{13}}{f_1 \mu_1} + \frac{K_{23}}{f_2 \mu_2}\right)\frac{1}{f_3 \mu_3}\right]\text{th}\,\sqrt{c - A_2 b}\,h - $$

$$\left. - A_2\left[\frac{h}{\mu_1} + \frac{1}{3}\frac{K_{13}h^3}{f_1 \mu_1 f_3 \mu_3}\right] - \left[\frac{h}{\mu_2} + \frac{1}{3}\frac{K_{23}h^3}{f_2 \mu_2 f_3 \mu_3}\right] + \frac{h}{c - A_2 b}\left(\frac{A_2 K_{13}}{f_1 \mu_1} + \frac{K_{23}}{f_2 \mu_2}\right)\frac{1}{f_3 \mu_3}\right\}\right]. \quad (24.19)$$

§ 25. Motion of Incompressible Viscous Three-Phase Media in a Circular Cylindrical Conduit

In order to consider the motion of viscous incompressible three-phase media, we direct the 0z axis along the conduit axis. We consider the motion axisymmetric. Then

$$v_{1r} = v_{2r} = v_{3r} = 0,$$

$$v_{1\varphi} = v_{2\varphi} = v_{3\varphi} = 0,$$

$$\frac{\partial v_{1r}}{\partial \varphi} = \frac{\partial v_{2r}}{\partial \varphi} = \frac{\partial v_{3r}}{\partial \varphi} = 0,$$

while the continuity equation (3.3) for each of the phases yields

$$\left. \begin{array}{l} \dfrac{\partial v_{1z}}{\partial z} = 0 \\[2mm] \dfrac{\partial v_{2z}}{\partial z} = 0 \\[2mm] \dfrac{\partial v_{3z}}{\partial z} = 0 \end{array} \right\} \tag{25.1}$$

In this case the system of differential equations of motion (8.3) can be written as

$$\left. \begin{array}{l} f_1 \mu_1 \left(\dfrac{d^2 v_{1z}}{dr^2} + \dfrac{1}{r} \dfrac{dv_{1z}}{dr} \right) + K_{21}(v_{2z} - v_{1z}) + K_{31}(v_{3z} - v_{1z}) = f_1 N \\[3mm] f_2 \mu_2 \left(\dfrac{d^2 v_{2z}}{dr^2} + \dfrac{1}{r} \dfrac{dv_{2z}}{dr} \right) + K_{21}(v_{1z} - v_{2z}) + K_{32}(v_{3z} - v_{2z}) = f_2 N \\[3mm] f_3 \mu_3 \left(\dfrac{d^2 v_{3z}}{dr^2} + \dfrac{1}{r} \dfrac{dv_{3z}}{dr} \right) + K_{13}(v_{1z} - v_{3z}) + K_{23}(v_{2z} - v_{3z}) = f_3 N \end{array} \right\} \tag{25.2}$$

Therefore, we have the system of equations (25.2) and the boundary conditions

$$\left. \begin{array}{l} v_{1z} = 0 \\ v_{2z} = 0 \\ v_{3z} = 0 \end{array} \right\} \tag{25.3}$$

at r = R for the determination of v_{1z}, v_{2z}, v_{3z}.

Adding the equations of the system (25.2) term by term, and integrating (25.3), we have

$$f_1 \mu_1 v_{1z} + f_2 \mu_2 v_{2s} + f_3 \mu_3 v_{3z} = \frac{N}{4} r^2 + c_1 \ln r + c_2.$$

Since ln r becomes infinite at r = 0, and the velocities of the media should be finite on the conduit axis, we set $c_1 = 0$ and determine c_2 from the boundary condition (25.3). Then the previous equation becomes

$$f_1 \mu_1 v_{1z} + f_2 \mu_2 v_{2z} + f_3 \mu_3 v_{3z} = \frac{N}{4} (r^2 - R^2). \tag{25.4}$$

We obtain the equations to determine v_{1z} and v_{2z} exactly the same as in § 24:

$$\frac{d^2 v_{1z}}{dr^2} + \frac{1}{r} \frac{dv_{1z}}{dr} + a v_{1z} + b v_{2z} = N \left[\frac{1}{\mu_1} - \frac{K_{13}}{4 f_1 \mu_1 f_3 \mu_3} (r^2 - R^2) \right], \tag{25.5}$$

$$\frac{d^2 v_{2z}}{dr^2} + \frac{1}{r} \frac{dv_{2z}}{dr} + c v_{2z} + d v_{1z} = N \left[\frac{1}{\mu_2} - \frac{K_{23}}{4 f_2 \mu_2 f_3 \mu_3} (r^2 - R^2) \right]. \tag{25.6}$$

Solving (25.5) and (25.6) by the d'Alembert method, we have

$$v_{1z} = - \frac{N}{A_2 - A_1} \left[\frac{1}{c - A_1 b} \left\{ \left[\left(\frac{A_1}{\mu_1} + \frac{1}{\mu_2} \right) - \frac{1}{c - A_1 b} \left(\frac{A_1 K_{13}}{f_1 \mu_1} + \frac{K_{23}}{f_2 \mu_2} \right) \frac{1}{f_3 \mu_3} \right] \frac{I_0 \left(\sqrt{c - A_1 b} \, r \right)}{I_0 \left(\sqrt{c - A_1 b} \, R \right)} - \right.$$

$$\left. - A_1 \left[\frac{1}{\mu_1} - \frac{K_{13}}{4 f_1 \mu_1 f_3 \mu_3} (r^2 - R^2) \right] - \left[\frac{1}{\mu_2} - \frac{K_{23}}{2 f_2 \mu_2 f_3 \mu_3} (r^2 - R^2) \right] + \frac{1}{c - A_1 b} \left[\frac{A_1 K_{13}}{f_1 \mu_1} + \frac{K_{23}}{f_2 \mu_2} \right] \frac{1}{f_3 \mu_3} \right\} - \frac{1}{c - A_2 b} \times$$

$$\times \left\{ \left[\left[\left(\frac{A_2}{\mu_1} + \frac{1}{\mu_2} \right) - \frac{1}{c - A_2 b} \left(\frac{A_2 K_{13}}{f_1 \mu_1} + \frac{K_{23}}{f_2 \mu_2} \right) \frac{1}{f_3 \mu_3} \right] \frac{I_0 \left(\sqrt{c - A_2 b} \, r \right)}{I_0 \left(\sqrt{c - A_2 b} \, R \right)} - A_2 \left[\frac{1}{\mu_1} - \frac{K_{13}}{4 \, f_1 \mu_1 f_3 \mu_3} \left(r^2 - R^2 \right) \right] - \right.$$

$$\left. - \left[\frac{1}{\mu_2} - \frac{K_{23}}{2 \, f_2 \mu_2 f_3 \mu_3} \left(r^2 - R^2 \right) \right] + \frac{1}{c - A_2 b} \left[\frac{A_2 K_{13}}{f_1 \mu_1} + \frac{K_{23}}{f_2 \mu_2} \right] \frac{1}{f_3 \mu_3} \right] \right\},$$

$$v_{2z} = - \frac{N}{A_1 - A_2} \left[\frac{A_2}{c - A_1 b} \left\{ \left[\left(\frac{A_1}{\mu_1} - \frac{1}{\mu_2} \right) + \frac{1}{c - A_1 b} \left(\frac{A_1 K_{13}}{f_1 \mu_1} + \frac{K_{23}}{f_2 \mu_2} \right) \frac{1}{f_3 \mu_3} \right] \frac{I_0 \left(\sqrt{c - A_1 b} \, r \right)}{I_0 \left(\sqrt{c - A_1 b} \, R \right)} - \right. \right.$$

$$\left. - A_1 \left[\frac{1}{\mu_1} - \frac{K_{13}}{4 \, f_1 \mu_1 f_3 \mu_3} \left(r^2 - R^2 \right) \right] - \left[\frac{1}{\mu_2} - \frac{K_{23}}{4 \, f_2 \mu_2 f_3 \mu_3} \left(r^2 - R^2 \right) \right] + \frac{1}{c - A_1 b} \left[\frac{A_1 K_{13}}{f_1 \mu_1} + \right. \right.$$

$$\left. + \frac{K_{23}}{f_2 \mu_2} \right] \frac{1}{f_3 \mu_3} \right\} - \frac{A_1}{c - A_2 b} \left\{ \left[\left(\frac{A_2}{\mu_1} + \frac{1}{\mu_2} \right) + \frac{1}{c - A_2 b} \left(\frac{A_2 K_{13}}{f_1 \mu_1} + \frac{K_{23}}{f_2 \mu_2} \right) \frac{1}{f_3 \mu_3} \right] \frac{I_0 \left(\sqrt{c - A_2 b} \, r}{I_0 \left(\sqrt{c - A_2 b} \, R} \right) - \right.$$

$$\left. \left. - A_2 \left[\frac{1}{\mu_1} - \frac{K_{13}}{4 \, f_1 \mu_1 f_3 \mu_3} \left(r^2 - R^2 \right) \right] - \left[\frac{1}{\mu_2} - \frac{K_{23}}{4 \, f_2 \mu_2 f_3 \mu_3} \left(r^2 - R^2 \right) \right] + \frac{1}{c - A_2 b} \left(\frac{A_2 K_{13}}{f_1 \mu_1} \cdot \frac{K_{23}}{f_2 \mu_2} \right) \frac{1}{f_3 \mu_3} \right\} \right],$$

$$v_{3z} = - \frac{1}{f_3 \mu_3} \frac{N}{A_1 - A_2} \left[\frac{f_1 \mu_1 + A_2 f_2 \mu_2}{c - A_1 b} \left\{ \left[\left(\frac{A_1}{\mu_1} - \frac{1}{\mu_2} \right) + \frac{1}{c - A_1 b} \left(\frac{A_1 K_{13}}{f_1 \mu_1} + \frac{K_{23}}{f_2 \mu_2} \right) \frac{1}{f_3 \mu_3} \right] \times \right. \right.$$

$$\times \frac{I_0 \left(\sqrt{c - A_1 b} \, r \right)}{I_0 \left(\sqrt{c - A_1 b} \, R \right)} + A_1 \left[\frac{1}{\mu_2} - \frac{K_{13}}{4 \, f_1 \mu_1 f_3 \mu_3} \left(r^2 - R^2 \right) \right] - \left[\frac{1}{\mu_2} - \frac{K_{23}}{4 \, f_2 \mu_2 f_3 \mu_3} \left(r^2 - R^2 \right) \right] + \frac{1}{c - A_1 b} \left(\frac{A_1 K_{13}}{f_1 \mu_1} + \right.$$

$$\left. + \frac{K_{23}}{f_2 \mu_2} \right) \frac{1}{f_3 \mu_3} \right\} - \frac{f_1 \mu_1 + A_1 f_2 \mu_2}{c - A_2 b} \left\{ \left[\frac{A_2}{\mu_1} - \frac{1}{\mu_2} + \frac{1}{c - A_1 b} \left(\frac{A_2 K_{13}}{f_1 \mu_1} + \frac{K_{23}}{f_2 \mu_2} \right) \frac{1}{f_3 \mu_3} \right] \frac{I_0 \left(\sqrt{c - A_2 b} \, r \right)}{I_0 \left(\sqrt{c - A_2 b} \, R \right)} - \right.$$

$$\left. \left. - A_2 \left[\frac{1}{\mu_1} - \frac{K_{13}}{4 \, f_1 \mu_1 f_3 \mu_3} \left(r^2 - R^2 \right) \right] - \left[\frac{1}{\mu_2} - \frac{K_{23}}{4 \, f_2 \mu_2 f_3 \mu_3} \left(r^2 - R^2 \right) \right] + \frac{1}{c - A_2 b} \left(\frac{A_2 K_{13}}{f_1 \mu_1} + \frac{K_{23}}{f_2 \mu_2} \right) \frac{1}{f_3 \mu_3} \right\} \right],$$

where

$$A_{1,2} = \frac{-(a - c) \pm \sqrt{(a - c)^2 + 4 \, bd}}{2b} \, .$$

In contrast to two-phase viscous flow, there are three interaction coefficients in the case under consideration. The velocity formulas obtained make it possible to determine quantitative values of the coefficients, and may also be applied in the investigation of some parameters of the mutually penetrating motion of three-phase viscous media, as for example in determining the optimal diameter, the drage coefficient of conduits, and the optimal concentration of the transportable mixture.

§ 26. Hydraulic Transport of Fine-Grained Materials of Heterogeneous Coarseness

The finest particles of soil and mineral resources which can be transported by hydraulic methods are actually heterogeneous in coarseness. As such particles move, the coarser fractions seem to displace the finer from their medium, with the result that the coarser particles sink to the bottom of the flow and the finer float in the upper layers [25]. The finer particles in the flow certainly affect the motion parameters of the coarser particles, and conversely. This all influences the energy capacity of the hydraulic transport. The question naturally arises as to which case has the smaller energy capacity; combined or individual transport of fine and coarse particles along the same conduit.

Let us consider the hydraulic transport of fine solid particles consisting of two fractions of differing coarseness [26]. Let the motion be between parallel walls extending to infinity in both directions. Let the 0x axis be along the axis of the plane conduit, and let the 0y axis be vertical to the 0x axis (Fig. 7). As is indicated above, an interface y = l will form in this case, with the first two-phase medium (water + fine particles) moving above the interface and the second two-phase medium (water + coarser particles) moving below it. We consider the motion to be laminar. We assume that there is no sinking or rising of particles in the flow; the motion is steady; the trajectories of all the particles are rectilinear and parallel to the 0x axis; the pressure for all the phases is identical, and the reduced densities are constant. Then by utilizing the equations of motion (6.2), (6.3), we have

$$\left.\begin{array}{l} f_1\mu_1 \dfrac{d^2u_1}{dy^2} + K_1(u_2 - u_1) = f_1 N \\[2mm] f_2\mu_2 \dfrac{d^2u_2}{dy^2} + K_1(u_1 - u_2) = f_2 N \end{array}\right\} \tag{26.1}$$

$$f_1 + f_2 = 1, \tag{26.2}$$

for the first two-phase medium, and

$$\left.\begin{array}{l} f_3\mu_1 \dfrac{d^2u_3}{\partial y^2} + K_2(u_4 - u_3) = f_3 N \\[2mm] f_4\mu_4 \dfrac{d^2u_4}{dy^2} + K_2(u_3 - u_4) = f_4 N \end{array}\right\} \tag{26.3}$$

$$f_3 + f_4 = 1, \tag{26.4}$$

for the second two-phase medium, where u_1, u_3 are the velocities of the water; u_2, u_4 are the velocities of the solid particles; $f_1 = \rho_1/\rho_{1i}$, $f_2 = \rho_2/\rho_{2i}$, $f_3 = \rho_3/\rho_{1i}$, $f_4 = \rho_4/\rho_{4i}$ are the porosities; K_1, K_2 the coefficients of media interaction; $N = \partial p/\partial x = \text{const}$ (let us note that $dp/\partial y = 0$); μ_1 the viscosity coefficient of water; and μ_2, μ_4 provisional coefficients of the solid particles.

The boundary conditions will be the following:

1. From the condition of adhesion of the media to the fixed bounding walls we have

$$u_1 = u_2 = 0 \tag{26.5}$$

for y = h (on the upper wall), and

$$u_3 = u_4 = 0 \tag{26.6}$$

for y = −h (on the lower wall);

2. On the interface y = l, evidently

$$u_1 = u_3; \tag{26.8}$$

in a particular case

$$\frac{du_2}{dy} = 0, \tag{26.8}$$

$$\frac{du_4}{dy} = 0, \tag{26.9}$$

and the tangential stresses of the mixture above and below y = l, should be equal, that is,

$$f_1\mu_1 \frac{du_1}{dy} + f_2\mu_2 \frac{du_2}{dy} = f_3\mu_1 \frac{du_3}{dy} + f_4\mu_4 \frac{du_4}{dy} \, .$$

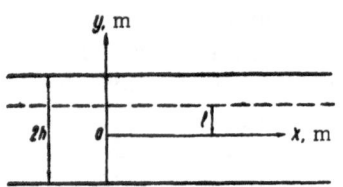

Fig. 7

Taking account of (26.8) and (26.9), we write the latter as

$$\left(\frac{du_3}{dy}\right)_{y=l} = \frac{f_1}{f_3}\left(\frac{du_1}{dy}\right)_{y=l} = \frac{f_1}{f_3}\,T. \tag{26.10}$$

Combining the equations in (26.1) term by term and taking account of (26.2), and then integrating twice with respect to y, we obtain

$$f_1\mu_1 u_1 + f_2\mu_2 u_2 = \frac{Ny^2}{2} + c_1 y + c_2. \tag{26.11}$$

Determining the constants of integration from the boundary conditions (26.5) and (26.10), we have

$$\left.\begin{array}{l} c_1 = f_1\mu_1 T - Nl \\ c_2 = (Nl - f_1\mu_1 T)\,h - \dfrac{Nh^2}{2} \end{array}\right\}. \tag{26.12}$$

On the basis of (26.11), the equations (26.1) are rewritten as

$$\frac{d^2 u_1}{dy^2} - K_1\left(\frac{1}{f_1\mu_1} + \frac{1}{f_2\mu_2}\right)u_1 = -\frac{K_1}{f_1\mu_1 f_2\mu_2}\left(\frac{Ny^2}{2} + c_1 y + c_2\right) + \frac{N}{\mu_1}\,, \tag{26.13}$$

$$\frac{d^2 u_2}{dy^2} - K_1\left(\frac{1}{f_1\mu_1} + \frac{1}{f_2\mu_2}\right)u_2 = -\frac{K_1}{f_1\mu_1 f_2\mu_2}\left(\frac{Ny^2}{2} + c_1 y + c_2\right) + \frac{N}{\mu_2}. \tag{26.14}$$

Reasoning analogously, we obtain for (26.3) and the boundary conditions (26.6), (26.9), and (26.10) when (26.4) is taken into account

$$\frac{d^2 u_3}{dy^2} - K_2\left(\frac{1}{f_3\mu_1} + \frac{1}{f_4\mu_4}\right)u_3 = -\frac{K_2}{f_3\mu_1 f_4\mu_4}\left(\frac{Ny^2}{2} + c_3 y + c_4\right) + \frac{N}{\mu_1}\,, \tag{26.15}$$

$$\frac{d^2 u_4}{dy^2} - K_2\left(\frac{1}{f_3\mu_1} + \frac{1}{f_4\mu_4}\right)u_4 = -\frac{K_2}{f_3\mu_1 f_4\mu_4}\left(\frac{Ny^2}{2} + c_3 y + c_4\right) + \frac{N}{\mu_1}\,, \tag{26.16}$$

where

$$\left.\begin{array}{l} c_3 = f_1\mu_1 T - Nl \\ c_4 = (f_1\mu_1 T - Nl)\,h - \dfrac{Nh^2}{2} \end{array}\right\}. \tag{26.17}$$

Therefore, the determination of the velocities of the media reduces to the solution of equations (26.13), (26.14), (26.15), and (26.16), under the appropriate boundary conditions (26.5), (26.6), (26.8), and (26.9). Such solutions will be

$$u_1 = -N\left\{\frac{1}{\sqrt{K\dfrac{f_1\mu_1 + f_2\mu_2}{f_1\mu_1 f_2\mu_2}}\;\mathrm{ch}\,\sqrt{K_1\dfrac{f_1\mu_1 + f_2\mu_2}{f_1\mu_1 f_2\mu_2}}\,(h-l)}\left[\left(\frac{1}{f_1\mu_1 + f_2\mu_2} - \frac{1}{\mu_1}\right)\frac{\mathrm{ch}\,\sqrt{K_1\dfrac{f_1\mu_1 + f_2\mu_2}{f_1\mu_1 f_2\mu_2}}\,(y-l)}{\sqrt{K_1\dfrac{f_1\mu_1 + f_2\mu_2}{f_1\mu_1 f_2\mu_2}}} - \right.\right.$$

$$\left.- \frac{f_2\mu_2 T}{N(f_1\mu_1 + f_2\mu_2)}\;\mathrm{sh}\,\sqrt{K_1\dfrac{f_1\mu_1 + f_2\mu_2}{f_1\mu_1 f_2\mu_2}}\,(y-h)\right] - \frac{f_1\mu_1 f_2\mu_2}{K_1(f_1\mu_1 + f_2\mu_2)}\left(\frac{1}{f_1\mu_1 + f_2\mu_2} - \frac{1}{\mu_1} - \right.$$

$$\left.\left.- \frac{K_1 h^2}{2 f_1\mu_1 f_2\mu_2} + \frac{K_1 lh}{f_1\mu_1 f_2\mu_2} - \frac{K_1 Th}{N f_2\mu_2}\right) + \frac{f_1\mu_1}{f_1\mu_1 + f_2\mu_2}\left(\frac{l}{f_1\mu_1} - \frac{T}{N}\right)y - \frac{1}{2(f_1\mu_1 + f_2\mu_2)}\,y^2\right\}\,, \tag{26.18}$$

$$u_2 = -N \left\{ \frac{1}{\sqrt{K_1 \frac{f_1\mu_1 + f_2\mu_2}{f_1\mu_1 f_2\mu_2}} \, \text{ch} \sqrt{K_1 \frac{f_1\mu_1 + f_2\mu_2}{f_1\mu_1 f_2\mu_2}} (h-l)} \left[\left(\frac{1}{f_1\mu_1 + f_2\mu_2} - \frac{1}{\mu_2} \right) \frac{\text{ch} \sqrt{K_1 \frac{f_1\mu_1 + f_2\mu_2}{f_1\mu_1 f_2\mu_2}} (y-l)}{\sqrt{K_1 \frac{f_1\mu_1 + f_2\mu_2}{f_1\mu_1 f_2\mu_2}}} + \right. \right.$$

$$\left. + \frac{f_1\mu_1 T}{N(f_1\mu_1 + f_2\mu_2)} \, \text{sh} \sqrt{K_1 \frac{f_1\mu_1 + f_2\mu_2}{f_1\mu_1 f_2\mu_2}} (y-h) \right] - \frac{f_1\mu_1 f_2\mu_2}{K_1 (f_1\mu_1 + f_2\mu_2)} \left(\frac{1}{f_1\mu_1 + f_2\mu_2} - \frac{1}{\mu_2} - \right.$$

$$\left. - \frac{K_1 h^2}{2 f_1\mu_1 f_2\mu_2} + \frac{K_1 lh}{f_1\mu_1 f_2\mu_2} - \frac{K_1 Th}{N f_2\mu_2} \right) + \frac{f_1\mu_1}{f_1\mu_1 + f_2\mu_2} \left(\frac{l}{f_1\mu_1} - \frac{T}{N} \right) y - \frac{1}{2(f_1\mu_1 + f_2\mu_2)} y^2 \right\}, \qquad (26.19)$$

$$u_3 = -N \left\{ \frac{1}{\sqrt{K_2 \frac{f_3\mu_1 + f_4\mu_4}{f_3\mu_1 f_4\mu_4}} \, \text{ch} \sqrt{K_2 \frac{f_3\mu_1 + f_4\mu_4}{f_3\mu_1 f_4\mu_4}} (h+l)} \left[\left(\frac{1}{f_3\mu_1 + f_4\mu_4} - \frac{1}{\mu_1} \right) \frac{\text{ch} \sqrt{K_2 \frac{f_3\mu_1 + f_4\mu_4}{f_3\mu_1 f_4\mu_4}} (y-l)}{\sqrt{K_2 \frac{f_3\mu_1 + f_4\mu_4}{f_3\mu_1 f_4\mu_4}}} - \right. \right.$$

$$\left. - \frac{f_1 f_4\mu_4 T}{N f_3 (f_3\mu_1 + f_4\mu_4)} \, \text{sh} \sqrt{K_2 \frac{f_3\mu_1 + f_4\mu_4}{f_3\mu_1 f_4\mu_4}} (y-h) \right] - \frac{f_3\mu_1 f_4\mu_4}{K_2 (f_3\mu_1 + f_4\mu_4)} \left(\frac{1}{f_3\mu_1 + f_4\mu_4} - \frac{1}{\mu_1} - \right.$$

$$\left. - \frac{K_2 h^2}{2 f_3\mu_1 f_4\mu_4} - \frac{K_2 lh}{f_3\mu_1 f_4\mu_4} + \frac{K_2 f_1 Th}{N f_3 f_4\mu_4} \right) + \frac{f_1\mu_1}{f_3\mu_1 + f_4\mu_4} \left(\frac{l}{f_1\mu_1} - \frac{T}{N} \right) y - \frac{1}{2(f_3\mu_1 + f_4\mu_4)} y^2 \right\}, \qquad (26.20)$$

$$u_4 = -N \left\{ \frac{1}{\sqrt{K_2 \frac{f_3\mu_1 + f_4\mu_4}{f_3\mu_1 f_4\mu_4}} \, \text{ch} \sqrt{K_2 \frac{f_3\mu_1 + f_4\mu_4}{f_3\mu_1 f_4\mu_4}} (h+l)} \left[\left(\frac{1}{f_3\mu_1 + f_4\mu_4} - \frac{1}{\mu_4} \right) \frac{\text{ch} \sqrt{K_2 \frac{f_3\mu_1 + f_4\mu_4}{f_3\mu_1 f_4\mu_4}} (y-l)}{\sqrt{K_2 \frac{f_3\mu_1 + f_4\mu_4}{f_3\mu_1 f_4\mu_4}}} + \right. \right.$$

$$\left. + \frac{f_1\mu_1 T}{N(f_3\mu_1 + f_4\mu_4)} \, \text{sh} \sqrt{K_2 \frac{f_3\mu_1 + f_4\mu_4}{f_3\mu_1 f_4\mu_4}} (y+h) \right] - \frac{f_3\mu_1 f_4\mu_4}{K_2 (f_3\mu_1 + f_4\mu_4)} \left(-\frac{1}{f_3\mu_1 + f_4\mu_4} - \frac{1}{\mu_4} - \frac{K_2 h^2}{2 f_3\mu_1 f_4\mu_4} - \right.$$

$$\left. - \frac{K_2 lh}{f_3\mu_1 f_4\mu_4} + \frac{K_2 f_1 Th}{N f_3 f_4\mu_4} \right) + \frac{f_1\mu_1}{f_3\mu_1 + f_4\mu_4} \left(\frac{l}{f_1\mu_1} - \frac{T}{N} \right) y - \frac{1}{2(f_3\mu_1 + f_4\mu_4)} y^2 \right\}. \qquad (26.21)$$

Furthermore, let us determine an expression for **T**. To do this, let us use the boundary conditions (26.7). Then the term T/N contained in the velocity formulas will be

$$\frac{T}{N} = \frac{A}{B}, \qquad (26.22)$$

where

$$A = \frac{\frac{1}{f_3\mu_1 + f_2\mu_2} - \frac{1}{\mu_1}}{K_1 \frac{f_1\mu_1 + f_2\mu_2}{f_1\mu_1 f_2\mu_2} \, \text{ch} \sqrt{K_1 \frac{f_1\mu_1 + f_2\mu_2}{f_1\mu_1 f_2\mu_2}} (h-l)} - \frac{\frac{1}{f_3\mu_1 + f_4\mu_4} - \frac{1}{\mu_1}}{K_2 \frac{f_3\mu_1 + f_4\mu_4}{f_3\mu_1 f_4\mu_4} \, \text{ch} \sqrt{K_2 \frac{f_3\mu_1 + f_4\mu_4}{f_3\mu_1 f_4\mu_4}} (h+l)} -$$

$$- \frac{f_1\mu_1 f_2\mu_2}{K_1 (f_1\mu_1 + f_2\mu_2)} \left(\frac{1}{f_1\mu_1 + f_2\mu_2} - \frac{1}{\mu_1} - \frac{K_1 h^2}{2 f_1\mu_1 f_2\mu_2} + \frac{K_1 lh}{f_1\mu_1 f_2\mu_2} \right) + \frac{f_3\mu_1 f_4\mu_4}{K_2 (f_3\mu_1 + f_4\mu_4)} \times$$

$$\times \left(\frac{1}{f_3\mu_1 + f_4\mu_4} - \frac{1}{\mu_1} - \frac{K_2 h^2}{2 f_3\mu_1 f_4\mu_4} - \frac{K_2 lh}{f_3\mu_1 f_4\mu_4} \right) + \frac{l^2}{2} \left(\frac{1}{f_1\mu_1 + f_2\mu_2} - \frac{1}{f_3\mu_1 + f_4\mu_4} \right); \qquad (26.23)$$

$$B = \cfrac{f_2\mu_2}{(f_1\mu_1 + f_2\mu_2)\sqrt{K_1 \dfrac{f_1\mu_1 + f_2\mu_2}{f_1\mu_1 f_2\mu_2}}\, \mathrm{ch}\sqrt{K_1 \dfrac{f_1\mu_1 + f_2\mu_2}{f_1\mu_1 f_2\mu_2}}(h-l)\ \mathrm{sh}\sqrt{K_1 \dfrac{f_1\mu_1 + f_2\mu_2}{f_1\mu_1 f_2\mu_2}}(l-h)} -$$

$$- \cfrac{f_1 f_4\mu_4}{f_3(f_3\mu_1 + f_4\mu_4)\sqrt{K_2 \dfrac{f_3\mu_1 + f_4\mu_4}{f_3\mu_1 f_4\mu_4}}\, \mathrm{ch}\sqrt{K_2 \dfrac{f_3\mu_1 + f_4\mu_4}{f_3\mu_1 f_4\mu_4}}(h+l)} \times$$

$$\times \mathrm{sh}\sqrt{K_2 \dfrac{f_3\mu_1 + f_4\mu_4}{f_3\mu_1 f_4\mu_4}}(l+h) - \left(\frac{1}{f_1\mu_1 + f_2\mu_2} + \frac{1}{f_3\mu_1 + f_4\mu_4}\right) f_1\mu_1 h + \left(\frac{1}{f_1\mu_1 + f_2\mu_2} - \frac{1}{f_3\mu_1 + f_4\mu_4}\right) f_1\mu_1 l. \quad (26.24)$$

Therefore, the velocities of the media in the case under consideration are distributed over the cross section according to formulas (26.18), (26.19), (26.20), and (26.21).

When velocities are present it is easy to derive expressions for the media discharges from the formulas

$$Q_1 = f_1 \int_l^h u_1 dy, \quad Q_2 = f_2 \int_l^h u_2 dy,$$

$$Q_3 = f_3 \int_{-h}^l u_3 dy, \quad Q_4 = f_4 \int_{-h}^l u_4 dy,$$

by substituting values of the velocities from (26.18), (26.19), (26.20), and (26.21) and integrating. We then obtain the following formulas for the discharges:

$$Q_1 = -Nf_1 L_1, \tag{26.25}$$

$$Q_2 = -Nf_2 L_2, \tag{26.26}$$

$$Q_3 = -Nf_3 L_3, \tag{26.27}$$

$$Q_4 = -Nf_4 L_4, \tag{26.28}$$

where

$$L_1 = \frac{1}{m_1\,\mathrm{ch}\sqrt{m_1}\,(h-l)}\left[\left(\frac{1}{f_1\mu_1 + f_2\mu_2} - \frac{1}{\mu_1}\right)\frac{\mathrm{sh}\sqrt{m_1}\,(h-l)}{\sqrt{m_1}} - \frac{f_2\mu_2 T\left(1-\mathrm{ch}\sqrt{m_1}\,(l-h)\right)}{N(f_1\mu_1 + f_2\mu_2)}\right] - \frac{1}{m_1}\left(\frac{1}{f_1\mu_1 + f_2\mu_2} - \frac{1}{\mu_1} - \right.$$
$$\left. - \frac{K_1 h^2}{2f_1\mu_1 f_2\mu_2} + \frac{K_1 lh}{f_2\mu_2 f_1\mu_1} - \frac{K_1 Th}{Nf_2\mu_2}\right)(h-l) + \frac{f_1\mu_1}{2(f_1\mu_1 + f_2\mu_2)}\left(\frac{l}{f_1\mu_1} - \frac{T}{N}\right)(h^2 - l^2) - \frac{h^3 - l^3}{6(f_1\mu_1 + f_2\mu_2)}; \tag{26.29}$$

$$L_2 = \frac{1}{m_1\,\mathrm{ch}\sqrt{m_1}\,(h-l)}\left[\left(\frac{1}{f_1\mu_1 + f_2\mu_2} - \frac{1}{\mu_2}\right)\frac{\mathrm{sh}\sqrt{m_1}(h-l)}{\sqrt{m_1}} + \frac{f_1\mu_1 T\left(1-\mathrm{ch}\sqrt{m_1}\,(l-h)\right)}{N(f_1\mu_1 + f_2\mu_2)}\right] - \frac{1}{m_1}\left(\frac{1}{f_1\mu_1 + f_2\mu_2} - \frac{1}{\mu_2} - \right.$$
$$\left. - \frac{K_1 h^2}{2f_1\mu_1 f_2\mu_2} + \frac{K_1 lh}{f_1\mu_1 f_2\mu_2} - \frac{K_1 Th}{Nf_2\mu_2}\right)(h-l) + \frac{f_1\mu_1}{2(f_1\mu_1 + f_2\mu_2)}\left(\frac{l}{f_1\mu_1} - \frac{T}{N}\right)(h^2 - l^2) - \frac{h^3 - l^3}{6(f_1\mu_1 + f_2\mu_2)}; \tag{26.30}$$

$$L_3 = \frac{1}{m_2\,\mathrm{ch}\sqrt{m_2}\,(h+l)}\left[\left(\frac{1}{f_2\mu_1 + f_4\mu_4} - \frac{1}{\mu_1}\right)\frac{\mathrm{sh}\sqrt{m_2}(h+l)}{\sqrt{m_2}} + \frac{f_1 f_4\mu_4 T\left(1-\mathrm{ch}\sqrt{m_2}\,(h+l)\right)}{Nf_3(f_3\mu_1 + f_4\mu_4)}\right] - \frac{1}{m_2}\left(\frac{1}{f_3\mu_1 + f_4\mu_4} - \frac{1}{\mu_1} - \right.$$
$$\left. - \frac{K_2 h^2}{2f_3\mu_1 f_4\mu_4} - \frac{K_2 lh}{f_3\mu_1 f_4\mu_4} + \frac{K_2 f_1 Th}{Nf_3 f_4\mu_4}\right)(h+l) + \frac{f_1\mu_1}{2(f_3\mu_1 + f_4\mu_4)}\left(\frac{l}{f_1\mu_1} - \frac{T}{N}\right)(l^2 - h^2) - \frac{h^3 + l^3}{6(f_3\mu_1 + f_4\mu_4)}; \tag{26.31}$$

$$L_4 = \frac{1}{m_2\,\mathrm{ch}\sqrt{m_2}\,(h+l)}\left[\left(\frac{1}{f_3\mu_1 + f_4\mu_4} - \frac{1}{\mu_4}\right)\frac{\mathrm{sh}\sqrt{m_2}\,(h+l)}{\sqrt{m_2}} - \frac{f_1\mu_1 T\left(1-\mathrm{ch}\sqrt{m_2}\,(h+l)\right)}{N(f_3\mu_1 + f_4\mu_4)}\right] - \frac{1}{m_2}\left(\frac{1}{f_3\mu_1 + f_4\mu_4} - \frac{1}{\mu_4} - \right.$$
$$\left. - \frac{K_2 h^2}{2f_3\mu_1 f_4\mu_4} - \frac{K_2 lh}{f_3\mu_1 f_4\mu_4} + \frac{K_2 f_1 Th}{Nf_3 f_4\mu_4}\right)(h+l) + \frac{f_1\mu_1}{2(f_3\mu_1 + f_4\mu_4)}\left(\frac{l}{f_1\mu_4} - \frac{T}{N}\right)(l^2 - h^2) - \frac{h^3 + l^3}{6(f_3\mu_1 + f_4\mu_4)}; \tag{26.32}$$

here

$$m_1 = K_1 \frac{f_1\mu_1 + f_2\mu_2}{f_1\mu_1 f_2\mu_2};$$

$$m_2 = K_2 \frac{f_3\mu_1 + f_4\mu_4}{f_3\mu_1 f_4\mu_4}.$$

Let us note that $Q_1 + Q_3$ and $Q_2 + Q_4$ are the discharges of the liquid and solid phases, respectively. Knowing experimental values of these discharges and of the other parameters in the above-mentioned formulas, the quantity l can be determined from the relation

$$\frac{Q_1 + Q_3}{Q_2 + Q_4} = F(l).$$

If the experimental values of all the discharges are known, as well as the velocities at two points, one of which is above and the other below $y = l$, then the unknown coefficients

$$\mu_2, \ \mu_4, \ K_1, \ K_2, \ l$$

can be found from the system of equations

$$\frac{Q_1}{Q_2} = F_1 \left(\mu_2, \ \mu_4, \ K_1, \ K_2, \ l\right),$$

$$\frac{Q_3}{Q_4} = F_2 \left(\mu_2, \ \mu_1, \ K_1, \ K_2, \ l\right),$$

$$\frac{Q_1}{Q_3} = F_3 \left(\mu_2, \ \mu_4, \ K_1, \ K_2, \ l\right),$$

$$\frac{u_1}{u_2} = F_4 \left(\mu_2, \ \mu_4, \ K_1, \ K_2, \ l\right),$$

$$\frac{u_3}{u_4} = F_5 \left(\mu_2, \ \mu_4, \ K_1, \ K_2, \ l\right) \ .$$

Let us examine the following particular cases:

1. Let $l = 0$, $\mu_2 = \mu_4$, $f_1 = f_3$, $f_2 = f_4$ (this case corresponds to the motion of two-phase media). Under these conditions the velocity and discharge formulas (13.8), (13.9), (13.15), and (13.16) are obtained as a particular case from the formulas presented above.

2. Let $l = 0$, $\mu_1 = \mu_2 = \mu_4$, $f_1 = f_3$, $f_2 = f_4$. Then we obtain the known formulas for a viscous incompressible fluid from the velocity and discharge formulas.

3. Let $l \neq 0$ and let the viscosities of the media be such that $\mu_1 = \mu_2 = \mu_3 = \mu_4$ can be assumed. Then the velocities will be different for identical viscosities.

On the basis of the discharge formulas obtained, the stream power may be investigated in order to compare the energy capacities.

Under the effect of a pressure drop per unit length $\partial p / \partial x$ in a plane conduit, let there be the motion of a mixture of two-phase media with the total discharge

$$Q_{mi} = Q_1 + Q_2 + Q_3 + Q_4.$$

In this case the power required to transport Q_{mi} a distance s is

$$N = - Q_{mi} \frac{\partial p}{\partial x} s.$$

If only the first two-phase medium is transported along the same conduit under the condition $Q_2' = 2Q_2$, or the second two-phase medium under the condition $Q_4' = 2Q_4$ (where Q_2', Q_4' are the discharges of the solid particles), then the powers required for the transport will be, respectively,

$$N_1 = - Q'_{1\,mi} \frac{\partial p'}{\partial x} s,$$

$$N_2 = - Q'_{2\,mi} \frac{\partial p''}{\partial x} s;$$

where

$$Q'_{1\,mi} = Q'_1 + 2Q_2,$$

$$Q'_{2\,mi} = Q'_2 + 2Q_4.$$

In order to determine which of the cases has the lower energy capacity, the combined or the separate transport of fine and coarse particles along the same conduit, it is necessary to evaluate $2N/(N_1 + N_2)$. If this ratio is greater than one, then separate transport is advantageous; if it is less than one, then combined transport is better.

Under the assumptions

$$Q'_1 = 2Q_1, \quad Q'_3 = 2Q_3$$

(these equalities are approximate), we have

$$\frac{2N}{N_1 + N_2} = \frac{(1 + M)\, \frac{\partial p}{\partial x}}{\frac{\partial p'}{\partial x} + M \frac{\partial p''}{\partial x}}, \tag{26.33}$$

where

$$M = \frac{Q_3 + Q_4}{Q_1 + Q_2} = \frac{f_3 L_3 + f_1 L_4}{f_1 L_1 + f_2 L_2}.$$

Evidently

$$\left.\begin{aligned}
\frac{\partial p}{\partial x} &= - \frac{Q_2}{L_2} \\[2mm]
\frac{\partial p'}{\partial x} &= - \frac{Q'_2}{M_1} = - \frac{2Q_2}{M_1} \\[2mm]
\frac{\partial p''}{\partial x} &= - \frac{Q'_4}{M_2} = - \frac{2Q_4}{M_2}
\end{aligned}\right\} \tag{26.34}$$

where

$$M_1 = f_2 \left\{ 2 \left(\frac{\mu_2}{f_1\mu_3 + f_2\mu_2} - 1 \right) \left[\frac{f_1 f_2 \mu_1}{K_1 (f_1\mu_1 + f_2\mu_2)} \right]^{3/2} \mu_2^{1/2} \times \right.$$
$$\left. \times \operatorname{th} \sqrt{ K_1 \frac{f_1\mu_1 + f_2\mu_2}{f_1\mu_1 f_2\mu_2} } \, h + \frac{2h^3}{3 (f_1\mu_1 + f_2\mu_2)} - \left(\frac{\mu_2}{f_1\mu_1 + f_2\mu_2} - 1 \right) \frac{2 f_1 f_2 \mu_1 h}{K_1 (f_1\mu_1 + f_2\mu_2)} \right\};$$

$$M_2 = f_4 \left\{ 2 \left(\frac{\mu_4}{f_3\mu_1 + f_4\mu_4} - 1 \right) \left[\frac{f_3 f_4 \mu_1}{K_2 (f_3\mu_1 + f_4\mu_4)} \right]^{3/2} \times \right.$$
$$\left. \times \mu_4^{1/2} \operatorname{th} \sqrt{ K_2 \frac{f_3\mu_1 + f_4\mu_4}{f_3\mu_1 f_4\mu_4} } \, h + \frac{2h^3}{3(f_3\mu_1 + f_4\mu_4)} - \left(\frac{\mu_4}{f_3\mu_1 + f_4\mu_4} - 1 \right) \frac{2 f_3 f_4 \mu_1 h}{K_2 (f_1\mu_1 + f_2\mu_2)} \right\}.$$

We have assumed that the coefficients $K_1\,K_2, \mu_2, \mu_4$ are identical in both kinds of motion considered above.

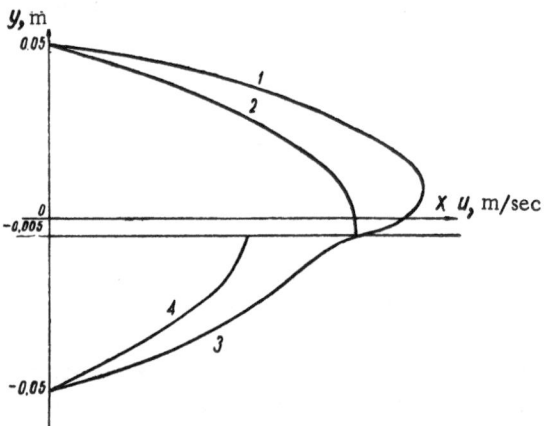

Fig. 8. Velocity distribution curves. 1, 3)
Fluid particles; 2, 4) finer and coarser
solid particles, respectively.

Substituting (26.34) into (26.33), we finally have after some manipulation

$$\frac{2N}{N_1 + N_2} = \frac{M_1 M_2 [f_1 L_1 + f_2 L_2 + f_3 L_3 + f_4 L_4]}{2 [L_2 M_2 f_2 (f_1 L_1 + f_2 L_2) + M_1 (f_3 L_3 + f_4 L_4) f_4 L_4]}. \tag{26.35}$$

In order to obtain numerical values for (26.35), it is necessary to clarify the nature of the interaction coefficients and the provisional viscosity coefficients experimentally as functions of the size of the fine particles.

K. V. Mukuk, in a dissertation prepared under the author's supervision, has experimentally shown the following: If the first phase is water and the second rosin particles, the provisional viscosity coefficient is greater for those homogeneous particles which have the greater size, i. e., in our case it is possible to assume $\mu_4 > \mu_2$. However, the relationship between the interaction coefficients has not yet been studied. An attempt has been made in [27] to obtain theoretical curves of the velocity distribution over the cross section of a plane conduit. For

$$\mu_1 = 10^{-4} \ \text{kg} \cdot \text{sec}/\text{m}^2, \ \mu_2 = 20 \cdot 10^{-4} \ \text{kg} \cdot \text{sec}/\text{m}^2,$$

$$K_1 = 1 \ \text{kg} \cdot \text{sec}/\text{m}^4 \quad \mu_4 = 30 \cdot 10^{-4} \ \text{kg} \cdot \text{sec}/\text{m}^2,$$

we have $u_1/u_2 = 1.4$ on the line $y = l + 0.01$ m, and $u_3/u_4 = 1.4$ on the line $y = l - 0.01$ m, and also we determine the quantities $K_2 = 1.832$ kg $\cdot$ sec/m^4, $l = 0.05$ m upon imposing the condition $u_1 = u_3 > u_2 > u_4$ on the interfacial line. Knowing the values of the parameters, we evaluate the values of the velocities on an electronic computer (Fig. 8).

CHAPTER IV

UNSTEADY MOTION OF VISCOUS TWO-PHASE MEDIA IN CONDUITS IN THE CONSTANT-POROSITY CASE

In the general case, investigations of the motion of incompressible two-phase media, all the motion parameters (porosities, pressures, velocities) should be considered time-dependent. Latipov [6, 28, 29] solved separate problems under the assumption of constant porosities and pressures independent of either the time or the coordinates.

§ 27. Pressureless Motion between Two Planes

Let a viscous incompressible two-phase medium filling the space between two horizontal infinite walls separated by a distance h be set in motion, starting at the time t = 0, by the movement of the upper wall to the right at the velocity $V(t)$, where $V(t) = V_0$ = const for $t \geq t_1$. We consider the pressure constant everywhere. We direct the 0x axis along the lower wall, which is considered fixed. The flow of the viscous two-phase medium will then be directed toward the right, parallel to the 0x axis. In this case the equations of motion (6.2), (6.3) are

$$\left.\begin{aligned}
\rho_1 \frac{\partial u_1}{\partial t} &= f_1 \mu_1 \frac{\partial^2 u_1}{\partial y^2} + K(u_2 - u_1) \\
\rho_2 \frac{\partial u_2}{\partial t} &= f_2 \mu_2 \frac{\partial^2 u_2}{\partial y^2} + K(u_1 - u_2)
\end{aligned}\right\} \tag{27.1}$$

The initial and boundary conditions will be

$$\left.\begin{aligned}
&\text{for } t = 0 & u_1 &= 0, \ u_2 = 0 \\
&\text{for } y = 0 & u_1 &= 0, \ u_2 = 0 \\
&\text{for } y = h & u_1 &= U(t), \ u_2 = U(t)
\end{aligned}\right\} \tag{27.2}$$

We apply the Laplace–Carson transformation in t to (27.1) and the boundary conditions (27.2) by means of the formulas

$$\bar{u}_1 = \lambda \int_0^\infty e^{-\lambda t} u_1 dt,$$

$$\bar{u}_2 = \lambda \int_0^\infty e^{-\lambda t} u_2 dt,$$

where λ is the transformation parameter. We then have

$$\frac{\partial u_1}{\partial t} \longrightarrow \div \longrightarrow \lambda \bar{u}_1, \quad \frac{\partial^2 u_1}{\partial y^2} \rightarrow \frac{d^2 \bar{u}_1}{dy^2},$$

$$\frac{\partial u_2}{\partial t} \longrightarrow \div \longrightarrow \lambda \bar{u}_2, \quad \frac{\partial^2 u_2}{\partial y^2} \rightarrow \frac{d^2 \bar{u}_2}{dy^2}.$$

Therefore, equations (27.1) and the initial and boundary conditions (27.2) transform into

$$\left.\begin{array}{l}\dfrac{d^2 \bar{u}_1}{dy^2} - \left(\dfrac{K}{f_1 \mu_1} + \dfrac{\lambda}{\nu_1}\right)\bar{u}_1 + \dfrac{K}{f_1 \mu_1}\,\bar{u}_2 = 0 \\[3mm] \dfrac{d^2 \bar{u}_2}{dy^2} - \left(\dfrac{K}{f_2 \mu_2} + \dfrac{\lambda}{\nu_2}\right)\bar{u}_2 + \dfrac{K}{f_2 \mu_2}\,\bar{u}_1 = 0\end{array}\right\} \tag{27.3}$$

and

$$\begin{array}{ll}\text{for } y = 0 & \bar{u}_1 = 0, \quad \bar{u}_2 = 0 \\[2mm] \text{for } y = h & \bar{u}_1 = U(\lambda), \quad \bar{u}_2 = U(\lambda)\end{array}\Biggr\} \tag{27.4}$$

where $\nu_1 = \mu_1/\rho_{1i}$, $\nu_2 = \mu_2/\rho_{2i}$ are the kinematic coefficients of viscosity of the two phases.

Solving the system (27.3) by the d'Alembert method under the boundary conditions (27.4), we have

$$\left.\begin{array}{l}\bar{u}_1 = \dfrac{\bar{U}(\lambda)}{A_1 - A_2}\left[(A_1 + 1)\dfrac{\text{sh}\,\sqrt{c - A_1 b}\,y}{\text{sh}\,\sqrt{c - A_1 b}\,h} - (A_2 + 1)\dfrac{\text{sh}\,\sqrt{c - A_2 b}\,y}{\text{sh}\,\sqrt{c - A_2 b}\,h}\right] \\[4mm] \bar{u}_2 = \dfrac{\bar{U}(\lambda)}{A_2 - A_1}\left[\left(A_2 - \dfrac{d}{b}\right)\dfrac{\text{sh}\,\sqrt{c - A_1 b}\,y}{\text{sh}\,\sqrt{c - A_1 b}\,h} - \left(A_1 - \dfrac{d}{b}\right)\dfrac{\text{sh}\,\sqrt{c - A_2 b}\,y}{\text{sh}\,\sqrt{c - A_2 b}\,h}\right]\end{array}\right\} \tag{27.5}$$

where

$$A_{1,2} = \frac{-(a - c) \pm \sqrt{(a - c)^2 + 4bd}}{2b} \tag{27.6}$$

is determined from

$$\frac{-Aa + d}{Ab - c} = A;$$

$$\left.\begin{array}{ll}a = \dfrac{K}{f_1 \mu_1} + \dfrac{\lambda}{\nu_1}, & b = \dfrac{K}{f_1 \mu_2} \\[3mm] c = \dfrac{K}{f_2 \mu_2} + \dfrac{\lambda}{\nu_2}, & d = \dfrac{K}{f_2 \mu_2}\end{array}\right\} \tag{27.7}$$

We invert the transform $\bar{u}_1$ to the original u_1 by means of the formula

$$u_1 = \frac{1}{2\pi i}\int_{\sigma - i\infty}^{\sigma + i\infty} \bar{u}_1\, e^{\lambda t}\frac{d\lambda}{\lambda}. \tag{27.8}$$

The function $F(\lambda) = \bar{u}_1/\lambda$ has two second-order branch points, which are the roots of the equation $A_1 - A_2 = 0$:

$$\lambda = \lambda_1 = \frac{-\left(\dfrac{1}{f_2 \mu_2} - \dfrac{1}{f_1 \mu_1}\right) + 2i\sqrt{\dfrac{1}{f_1 \mu_1} \cdot \dfrac{1}{f_2 \mu_2}}}{\dfrac{1}{\nu_2} - \dfrac{1}{\nu_1}} K,.$$

$$\lambda = \lambda_2 = \frac{-\left(\dfrac{1}{f_2 \mu_2} - \dfrac{1}{f_1 \mu_1}\right) - 2i\sqrt{\dfrac{1}{f_1 \mu_1} \cdot \dfrac{1}{f_2 \mu_2}}}{\dfrac{1}{\nu_2} - \dfrac{1}{\nu_1}} K$$

and the simple poles

$$\lambda = \lambda_0 = 0,$$
$$\lambda = \lambda_{1n} = -\gamma_{1n},$$
$$\lambda = \lambda_{2n} = -\gamma_{2n},$$

where λ_{1n}, λ_{2n} are the roots of the equations

$$c - A_1 b = -\frac{\pi^2 n^2}{h^2},$$
$$c - A_2 b = -\frac{\pi^2 n^2}{h^2},$$

corresponding to the expression

$$\gamma_{1n,\,2n} = \frac{1}{2\nu_1 \nu_2}\left[\frac{K}{\nu_1 \nu_2}\left(\frac{1}{\rho_1} + \frac{1}{\rho_2}\right) + \frac{\pi^2 n^2}{h^2}\left(\frac{1}{\nu_2} + \frac{1}{\nu_1}\right) \mp \right.$$
$$\mp \left.\left\{\left[\frac{K}{\nu_1 \nu_2}\left(\frac{1}{\rho_1} + \frac{1}{\rho_2}\right)\right]^2 + \frac{K}{\nu_1 \nu_2}\left(\frac{1}{\rho_1} - \frac{1}{\rho_2}\right)\left(\frac{1}{\nu_1} - \frac{1}{\nu_2}\right)\frac{\pi^2 n^2}{h^2} + \left[\frac{\pi^2 n^2}{h^2}\left(\frac{1}{\nu_2} - \frac{1}{\nu_1}\right)\right]^2\right\}^{1/2}\right].$$

We determine the residues by the customary method

$$c_0 = \frac{y}{h},$$
$$c_1 = 0,$$
$$c_2 = 0,$$
$$c_{1n} = (-1)^{n+1}\frac{R_1}{R_2}\frac{4\pi n}{h^2}\sin\left(\pi n\,\frac{y}{h}\right)\frac{U}{\gamma_{1n}},$$
$$c_{2n} = (-1)^{n+1}\frac{R_3}{R_4}\frac{4\pi n}{h^2}\sin\left(\pi n\,\frac{y}{h}\right)\frac{U}{\gamma_{2n}},$$

where

$$R_1 = \frac{\left(\dfrac{1}{f_1 \mu_1} + \dfrac{1}{f_2 \mu_2}\right)K - \dfrac{\gamma_{1n}}{\nu_2} + \dfrac{\pi^2 n^2}{h^2}}{\left(\dfrac{1}{f_1 \mu_1} + \dfrac{1}{f_2 \mu_2}\right)K - \left(\dfrac{1}{\nu_1} + \dfrac{1}{\nu_2}\right)\gamma_{1n} + 2\,\dfrac{\pi^2 n^2}{h^2}};$$

$$R_2 = \frac{\left(\dfrac{1}{\nu_2} - \dfrac{1}{\nu_1}\right)\left[\left(\dfrac{1}{f_2 \mu_2} - \dfrac{1}{f_1 \mu_1}\right)K - \gamma_{1n}\left(\dfrac{1}{\nu_2} - \dfrac{1}{\nu_1}\right)\right]}{\left(\dfrac{1}{f_1 \mu_1} + \dfrac{1}{f_2 \mu_2}\right)K - \left(\dfrac{1}{\nu_1} + \dfrac{1}{\nu_2}\right)\gamma_{1n} + 2\,\dfrac{\pi^2 n^2}{h^2}} - \left(\dfrac{1}{\nu_1} + \dfrac{1}{\nu_2}\right);$$

$$R_3 = \frac{\left(\dfrac{1}{f_1\mu_1} + \dfrac{1}{f_2\mu_2}\right)K - \dfrac{\gamma_{2n}}{\nu_2} + \dfrac{\pi^2 n^2}{h^2}}{\left(\dfrac{1}{f_1\mu_1} + \dfrac{1}{f_2\mu_2}\right)K - \left(\dfrac{1}{\nu_1} + \dfrac{1}{\nu_2}\right)\gamma_{2n} + 2\dfrac{\pi^2 n^2}{h^2}} \; ;$$

$$R_4 = \frac{\left(\dfrac{1}{\nu_2} - \dfrac{1}{\nu_1}\right)\left[\left(\dfrac{1}{f_2\mu_2} - \dfrac{1}{f_1\mu_1}\right)K - \gamma_{2n}\left(\dfrac{1}{\nu_2} - \dfrac{1}{\nu_1}\right)\right]}{\left(\dfrac{1}{f_1\mu_1} + \dfrac{1}{f_2\mu_2}\right)K - \left(\dfrac{1}{\nu_1} + \dfrac{1}{\nu_2}\right)\gamma_{1n} + 2\dfrac{\pi^2 n^2}{h^2}} - \left(\dfrac{1}{\nu_1} + \dfrac{1}{\nu_2}\right).$$

We used the Jordan lemma in evaluating the integral (27.8). Evaluating the integrals in small neighborhoods around the singular points and below and above the edges of the slits at the branch points, and also taking account of the well-known formulas [8]

$$\frac{1}{2\pi i}\int_{\sigma-i\infty}^{\sigma+i\infty} e^{\lambda t}\frac{d\lambda}{\lambda} = 1, \quad \frac{1}{2\pi i}\int_{\sigma-i\infty}^{\sigma+i\infty} e^{\lambda t}\frac{d\lambda}{\lambda+\gamma_{1n}} = e^{-\gamma_{1n}t}, \quad \frac{1}{2\pi i}\int_{\sigma-i\infty}^{\sigma+i\infty} e^{\lambda t}\frac{d\lambda}{\lambda+\gamma_{2n}} = e^{-\gamma_{2n}t},$$

we obtain the final formula for u_1:

$$u_1 = U(0)\frac{y}{h} + \sum_{k=1}^{s}\frac{U_k}{A_{1k}-A_{2k}}\left[(A_{1k}+1)\frac{\mathrm{sh}\sqrt{c-A_{1k}b}\,y}{\mathrm{sh}\sqrt{c-A_{1k}b}\,h}\right.$$

$$\left.-(A_{2k}+1)\frac{\mathrm{sh}\sqrt{c-A_{2k}b}\,y}{\mathrm{sh}\sqrt{c-A_{2k}b}\,h}\right]\frac{1}{\alpha_k}e^{-\alpha_k t} + \sum_{n=1}^{\infty}(-1)^{n+1}\frac{4\pi n}{h^2}\sin\left(\pi n\frac{y}{h}\right)\times$$

$$\times\left\{\frac{\dfrac{\left(\dfrac{1}{\rho_2\nu_2}+\dfrac{1}{\rho_1\nu_1}\right)K + \dfrac{\pi^2 n^2}{h^2} - \dfrac{\gamma_{1n}}{\nu_2}}{\left(\dfrac{1}{\rho_2\nu_2}+\dfrac{1}{\rho_1\nu_1}\right)K + 2\dfrac{\pi^2 n^2}{h^2} - \left(\dfrac{1}{\nu_2}+\dfrac{1}{\nu_1}\right)\gamma_{1n}}}{\dfrac{\left(\dfrac{1}{\nu_2}-\dfrac{1}{\nu_1}\right)\left[\left(\dfrac{1}{\rho_2\nu_2}-\dfrac{1}{\rho_1\nu_1}\right)K - \gamma_{1n}\left(\dfrac{1}{\nu_2}-\dfrac{1}{\nu_1}\right)\right]}{\left(\dfrac{1}{\rho_2\nu_2}+\dfrac{1}{\rho_1\nu_1}\right)K + 2\dfrac{\pi^2 n^2}{h^2} - \left(\dfrac{1}{\nu_2}+\dfrac{1}{\nu_1}\right)\gamma_{1n}} - \left(\dfrac{1}{\nu_2}+\dfrac{1}{\nu_1}\right)}\frac{U(\gamma_{1n})}{\gamma_{1n}}\cdot e^{-\gamma_{1n}t}\right.$$

$$\left.-\frac{\dfrac{\left(\dfrac{1}{\rho_2\nu_2}+\dfrac{1}{\rho_1\nu_1}\right)K + \dfrac{\pi^2 n^2}{h^2} - \dfrac{\gamma_{2n}}{\nu_2}}{\left(\dfrac{1}{\rho_2\nu_2}+\dfrac{1}{\rho_1\nu_1}\right)K + 2\dfrac{\pi^2 n^2}{h^2} - \left(\dfrac{1}{\nu_1}+\dfrac{1}{\nu_2}\right)\gamma_{2n}}}{\dfrac{\left[\left(\dfrac{1}{\rho_2\nu_2}-\dfrac{1}{\rho_1\nu_1}\right)K - \left(\dfrac{1}{\nu_2}-\dfrac{1}{\nu_1}\right)\gamma_{2n}\right]\left(\dfrac{1}{\nu_2}-\dfrac{1}{\nu_1}\right)}{\left(\dfrac{1}{\rho_2\nu_2}+\dfrac{1}{\rho_1\nu_1}\right)K + 2\dfrac{\pi^2 n^2}{h^2} - \left(\dfrac{1}{\nu_1}+\dfrac{1}{\nu_2}\right)\gamma_{2n}} - \left(\dfrac{1}{\nu_1}+\dfrac{1}{\nu_2}\right)}\frac{U(\gamma_{2n})}{\gamma_{2n}}e^{-\gamma_{2n}t}\right\}. \qquad (27.9)$$

Completely analogously we have

$$u_2 = U(0)\frac{y}{h} + \sum_{k=1}^{s}\frac{U_k}{A_{1k}-A_{2k}}\left[\left(A_{2k}-\frac{d}{b}\right)\frac{\mathrm{sh}\sqrt{c-A_{1k}b}\,y}{\mathrm{sh}\sqrt{c-A_{1k}b}\,h}\right.$$

$$\left.-\left(A_{1k}-\frac{d}{b}\right)\frac{\mathrm{sh}\sqrt{c-A_{2k}b}\,y}{\mathrm{sh}\sqrt{c-A_{2k}b}\,h}\right]\frac{1}{\alpha_k}e^{-\alpha_k t} + \sum_{n=1}^{\infty}(-1)^{n+1}\frac{4\pi n}{h^3}\sin\left(\pi n\frac{y}{h}\right)\times$$

$$\times \left\{ \frac{\dfrac{\left(\dfrac{1}{\rho_1 \nu_1} + \dfrac{1}{\rho_2 \nu_2}\right)K + \dfrac{\pi^2 n^2}{h^2} - \dfrac{1}{\nu_1}\gamma_{1n}}{\dfrac{\left(\dfrac{1}{\rho_2 \nu_2} + \dfrac{1}{\rho_1 \nu_1}\right)K + 2\dfrac{\pi^2 n^2}{h^2} - \left(\dfrac{1}{\nu_1} + \dfrac{1}{\nu_2}\right)\gamma_{1n}}{\dfrac{\left(\dfrac{1}{\nu_2} - \dfrac{1}{\nu_1}\right)\left[\left(\dfrac{1}{\rho_2 \nu_2} - \dfrac{1}{\rho_1 \nu_1}\right)K - \left(\dfrac{1}{\nu_2} - \dfrac{1}{\nu_1}\right)\gamma_{1n}\right]}{\left(\dfrac{1}{\rho_2 \nu_2} + \dfrac{1}{\rho_1 \nu_1}\right)K + 2\dfrac{\pi^2 n^2}{h^2} - \left(\dfrac{1}{\nu_1} + \dfrac{1}{\nu_2}\right)\gamma_{1n}} - \left(\dfrac{1}{\nu_1} + \dfrac{1}{\nu_2}\right)}} \frac{U(\gamma_{1n})}{\gamma_{1n}} e^{-\gamma_{1n}t} - \right.$$

$$\left. - \frac{\dfrac{\left(\dfrac{1}{\rho_2 \nu_2} + \dfrac{1}{\rho_1 \nu_1}\right)K + \dfrac{\pi^2 n^2}{h^2} - \dfrac{\gamma_{2n}}{\nu_1}}{\dfrac{\left(\dfrac{1}{\rho_2 \nu_2} + \dfrac{1}{\rho_1 \nu_1}\right)K + 2\dfrac{\pi^2 n^2}{h^2} - \left(\dfrac{1}{\nu_1} + \dfrac{1}{\nu_2}\right)\gamma_{2n}}{\dfrac{\left(\dfrac{1}{\nu_2} - \dfrac{1}{\nu_1}\right)\left[\left(\dfrac{1}{\rho_2 \nu_2} - \dfrac{1}{\rho_1 \nu_1}\right)K - \gamma_{2n}\left(\dfrac{1}{\nu_2} - \dfrac{1}{\nu_1}\right)\right]}{\left(\dfrac{1}{\rho_2 \nu_2} + \dfrac{1}{\rho_1 \nu_1}\right)K + 2\dfrac{\pi^2 n^2}{h^2} - \left(\dfrac{1}{\nu_1} + \dfrac{1}{\nu_2}\right)\gamma_{2n}} - \left(\dfrac{1}{\nu_1} - \dfrac{1}{\nu_2}\right)}} \frac{U(\gamma_{2n})}{\gamma_{2n}} e^{-\gamma_{2n}t} \right\}. \tag{27.10}$$

Since the equality

$$u_1 = u_2 = U(t)$$

is valid for y = h, we obtain from the preceding expressions

$$U(t) = U(0) + \sum_{k=1}^{s} \frac{c_k}{\alpha_k} e^{-\alpha_k t} .$$

Let us determine the tangential stresses since they are of importance to our subsequent investigations. Let τ_1 and τ_2 be the friction forces originating because of the first- and second-phase motions. Then the friction τ acting on some surface will be

$$\tau = \tau_1 + \tau_2.$$

Since the first phase acts on the fraction f_1, and the second on the fraction f_2 of this surface, we have on the basis of the Newton hypothesis

$$\tau_1 = f_1 \mu_1 \frac{\partial u_1}{\partial y},$$

$$\tau_2 = f_2 \mu_2 \frac{\partial u_2}{\partial y}.$$

Therefore, we obtain for the frictional force per unit surface

$$\tau = f_1 \mu_1 \frac{\partial u_1}{\partial y} + f_2 \mu_2 \frac{\partial u_2}{\partial y}. \tag{27.11}$$

Substituting the expressions for u_1 and u_2 from (27.9) and (27.10) into (27.11), we determine the formula for the friction stresses in the form

$$\tau = \frac{U(0)}{h}\left(\frac{\rho_1}{\rho_{1i}}\mu_1 + \frac{\rho_2}{\rho_{2i}}\mu_2\right) + \sum_{k=1}^{s} \frac{U_k}{A_{1k} - A_{2k}}\left\{\sqrt{(c - A_{1k}b)}\left[\frac{\rho_1}{\rho_{1i}}\mu_1(A_{1k} + 1) - \right.\right.$$

$$\left.\left. - \frac{\rho_2}{\rho_{2i}}\mu_2\left(A_{2k} - \frac{d}{b}\right)\right]\frac{\text{ch}\sqrt{(c - A_{1k}b)}\,y}{\text{sh}\sqrt{(c - A_{1k}b)}\,h} - \sqrt{(c - A_{2k}b)}\left[\frac{\rho_1}{\rho_{1i}}\mu_1(A_{2k} + 1) - \frac{\rho_2}{\rho_{2i}}\mu_2\left(A_{1k} - \frac{d}{b}\right)\right]\times\right.$$

$$\times \frac{\mathrm{ch}\sqrt{(c-A_{2k}b)}\,y}{\mathrm{sh}\sqrt{(c-A_{2k}b)}\,h}\Bigg\}\frac{1}{\alpha_k}e^{-\alpha_k t} + \sum_{n=1}^{\infty}\frac{4\pi^2}{h^2}(-1)^{n+1}n^2\cos\left(\pi n\,\frac{y}{h}\right)\times$$

$$\times\Bigg\{ \frac{\left[\left(\frac{\rho_{2i}}{\rho_2\mu_2}+\frac{\rho_{1i}}{\rho_1\mu_1}\right)K+\frac{\pi^2 n^2}{h^2}\right]\left(\frac{\rho_1}{\rho_{1i}}\mu_1+\frac{\rho_2}{\rho_{2i}}\mu_2\right)-\left(\frac{\rho_1}{\rho_{1i}}\frac{\rho_{2i}}{\mu_2}\mu_1+\frac{\rho_2}{\rho_{2i}}\frac{\rho_{1i}}{\mu_1}\mu_2\right)\gamma_{1n}}{\dfrac{\left(\frac{\rho_{2i}}{\rho_2\mu_2}+\frac{\rho_{1i}}{\rho_1\mu_1}\right)K+2\frac{\pi^2 n^2}{h^2}-\left(\frac{\rho_{2i}}{\mu_2}+\frac{\rho_{1i}}{\mu_1}\right)\gamma_{1n}}{\dfrac{\left(\frac{\rho_{2i}}{\mu_2}-\frac{\rho_{1i}}{\mu_1}\right)\left[\left(\frac{\rho_{2i}}{\rho_2\mu_2}-\frac{\rho_{1i}}{\rho_1\mu_1}\right)K-\gamma_{2n}\left(\frac{1}{\nu_2}-\frac{1}{\nu_1}\right)\right]}{\left(\frac{\rho_{2i}}{\rho_2\mu_2}+\frac{\rho_{1i}}{\rho_1\mu_1}\right)K+2\frac{\pi^2 n^2}{h^2}-\left(\frac{\rho_{2i}}{\mu_2}+\frac{\rho_{1i}}{\mu_1}\right)\gamma_{1n}}-\left(\frac{\rho_{2i}}{\mu_2}+\frac{\rho_{1i}}{\mu_1}\right)}}\cdot\frac{U(\gamma_{1n})}{\gamma_{1n}}e^{-\gamma_{1n}t}-$$

$$-\frac{\left[\left(\frac{\rho_{2i}}{\rho_2\mu_2}+\frac{\rho_{1i}}{\rho_1\mu_1}\right)K+\frac{\pi^2 n^2}{h^2}\right]\left(\frac{\rho_1}{\rho_{1i}}\mu_1+\frac{\rho_2}{\rho_{2i}}\mu_2\right)-\left(\frac{\rho_1}{\rho_{1i}}\frac{\rho_{2i}}{\mu_2}\mu_1+\frac{\rho_2}{\rho_{2i}}\frac{\rho_{1i}}{\mu_1}\mu_2\right)\gamma_{2n}}{\dfrac{\left(\frac{\rho_{2i}}{\rho_2\mu_2}+\frac{\rho_{1i}}{\rho_1\mu_1}\right)K+2\frac{\pi^2 n^2}{h^2}-\left(\frac{\rho_{2i}}{\mu_2}+\frac{\rho_{1i}}{\mu_1}\right)\gamma_{2n}}{\dfrac{\left(\frac{\rho_{2i}}{\mu_2}-\frac{\rho_{1i}}{\mu_1}\right)\left[\left(\frac{\rho_{2i}}{\rho_2\mu_2}-\frac{\rho_{1i}}{\rho_1\mu_1}\right)K-\gamma_{2n}\left(\frac{1}{\nu_2}-\frac{1}{\nu_1}\right)\right]}{\left(\frac{\rho_{2i}}{\rho_2\mu_2}+\frac{\rho_{1i}}{\rho_1\mu_1}\right)K+2\frac{\pi^2 n^2}{h^2}-\left(\frac{\rho_{2i}}{\mu_2}+\frac{\rho_{1i}}{\mu_1}\right)\gamma_{2n}}-\left(\frac{\rho_{2i}}{\mu_2}+\frac{\rho_{1i}}{\mu_1}\right)}}\cdot\frac{U(\gamma_{2n})}{\gamma_{2n}}e^{-\gamma_{2n}t}\Bigg\}. \qquad (27.12)$$

When $\mu_1=\mu_2=\mu$ and $\rho_{1i}=\rho_{2i}=\rho$, the formulas for the rates of the friction stresses take the form of the formulas for a single-phase flow [10].

§ 28. Pressure Motions between Fixed Planes

Let us consider the problem when the motion in a plane conduit (with walls separated by a distance 2h) is in the direction of the 0x axis, drawn down the middle between the walls, because of a pressure drop. Let us idealize the problem somewhat: the two-phase flow is at rest at the initial instant. After the motion has begun (t > 0), the pressure drop per unit time at once takes on the constant value N. Then the initial and boundary conditions will evidently be

$$\left.\begin{array}{lll}\text{for t} = 0 & u_1=0, & u_2=0 \\[4pt] \text{for } y=h \ (t>0) & u_1=0, & u_2=0 \\[4pt] \text{for } y=-h \ (t>0) & u_1=0, & u_2=0\end{array}\right\} \qquad (28.1)$$

Let us apply the Laplace–Carson transformation to the equations of motion (27.1), in which we take account of the pressure drop. We then obtain

$$\left.\begin{array}{l}\dfrac{d^2\bar{u}_1}{dy^2}-\left(\dfrac{K}{f_1\mu_1}+\dfrac{\lambda}{\nu_1}\right)\bar{u}_1+\dfrac{1}{f_1\mu_1}\bar{u}_2=\dfrac{1}{\mu_1}N \\[14pt] \dfrac{d^2\bar{u}_2}{dy^2}-\left(\dfrac{K}{f_2\mu_2}+\dfrac{\lambda}{\nu_2}\right)\bar{u}_2+\dfrac{1}{f_2\mu_2}\bar{u}_1=\dfrac{1}{\mu_2}N\end{array}\right\} \qquad (28.2)$$

Conditions (28.1) now become

$$\left.\begin{array}{lll}\text{for } y=h & \bar{u}_1=0, & \bar{u}_2=0 \\[4pt] \text{for } y=-h & \bar{u}_1=0, & \bar{u}_2=0\end{array}\right\} \qquad (28.3)$$

The problem is therefore reduced to solving the system of ordinary differential equations (28.2) under the boundary conditions (28.3). Solving this system and returning to the originals, we have the following formulas for the velocities:

$$u_1 = -\frac{\frac{p_{1i}}{p_1\mu_1}\frac{p_{2i}}{p_2\mu_2}}{\frac{p_{1i}}{p_1\mu_1}+\frac{p_{2i}}{p_2\mu_2}} N\left\{-\frac{1}{2}(y^2-h^2)-\frac{\frac{1}{\mu_1}-\frac{1}{\mu_2}}{\left(\frac{p_{1i}}{p_1\mu_1}+\frac{p_{2i}}{p_2\mu_2}\right)K}\frac{p_2\mu_2}{p_{2i}}\left[1-\frac{\operatorname{ch}\sqrt{\left(\frac{p_{1i}}{p_1\mu_1}+\frac{p_{2i}}{p_2\mu_2}\right)Ky}}{\operatorname{ch}\sqrt{\left(\frac{p_{1i}}{p_1\mu_1}+\frac{p_{2i}}{p_2\mu_2}\right)Kh}}\right]\right\}+$$

$$+\frac{4}{\pi}\sum_{n=1}^{\infty}\frac{(-1)^{n+1}}{\left(n+\frac{1}{2}\right)}N\cos\pi\left(n+\frac{1}{2}\right)\frac{y}{h}\left\{\frac{D_1}{D_2}\frac{1}{\gamma_{1n}}e^{-\gamma_{1n}t}-\frac{D_3}{D_4}\frac{1}{\gamma_{2n}}e^{-\gamma_{2n}t}\right\},$$

where

$$D_1 = \frac{\left[-\frac{p_{2i}}{\mu_2}\gamma_{1n}+\frac{p_{2i}}{p_2\mu_2}K+\frac{\pi^2}{h^2}\left(n+\frac{1}{2}\right)^2\right]\frac{1}{\mu_1}+\frac{p_{1i}}{p_1\mu_1}K\frac{1}{\mu_2}}{-\left(\frac{p_{2i}}{\mu_2}+\frac{p_{1i}}{\mu_1}\right)\gamma_{1n}+\left(\frac{p_{2i}}{p_2\mu_2}+\frac{p_{1i}}{p_1\mu_1}\right)K+2\frac{\pi^2}{h^2}\left(n+\frac{1}{2}\right)^2};$$

$$D_2 = -\left(\frac{p_{2i}}{\mu_2}+\frac{p_{1i}}{\mu_1}\right)+\frac{\left(\frac{p_{2i}}{\mu_2}-\frac{p_{1i}}{\mu_1}\right)\left[\left(\frac{p_{2i}}{p_2\mu_2}-\frac{p_{1i}}{p_1\mu_1}\right)K-\gamma_{1n}\left(\frac{1}{\nu_2}-\frac{1}{\nu_1}\right)\right]}{-\left(\frac{p_{2i}}{\mu_2}+\frac{p_{1i}}{\mu_1}\right)\gamma_{1n}-\left(\frac{p_{2i}}{p_2\mu_2}+\frac{p_{1i}}{p_1\mu_1}\right)K+2\frac{\pi^2}{h^2}\left(n+\frac{1}{2}\right)^2};$$

$$D_3 = \frac{\left[-\frac{p_{2i}}{\mu_2}\gamma_{2n}+\frac{p_{2i}}{p_2\mu_2}K+\frac{\pi^2}{h^2}\left(n+\frac{1}{2}\right)^2\right]\frac{1}{\mu_1}+\frac{p_{1i}}{p_1\mu_1}K\frac{1}{\mu_2}}{-\left(\frac{p_{2i}}{\mu_2}+\frac{p_{1i}}{\mu_1}\right)\gamma_{2n}+\left(\frac{p_{2i}}{p_2\mu_2}+\frac{p_{1i}}{p_1\mu_1}\right)K+2\frac{\pi^2}{h^2}\left(n+\frac{1}{2}\right)^2};$$

$$D_4 = -\left(\frac{p_{2i}}{\mu_2}+\frac{p_{1i}}{\mu_1}\right)+\frac{\left(\frac{p_{2i}}{\mu_2}-\frac{p_{1i}}{\mu_1}\right)\left[\left(\frac{p_{2i}}{p_2\mu_2}-\frac{p_{1i}}{p_1\mu_1}\right)K-\gamma_{2n}\left(\frac{1}{\nu_2}-\frac{1}{\nu_1}\right)\right]}{-\left(\frac{p_{2i}}{\mu_2}+\frac{p_{1i}}{\mu_1}\right)\gamma_{1n}+\left(\frac{p_{2i}}{p_2\mu_2}+\frac{p_{1i}}{p_1\mu_1}\right)K+2\frac{\pi^2}{h^2}\left(n+\frac{1}{2}\right)^2};$$

$$\gamma_{1n,2n} = \frac{1}{2}\nu_1\nu_2\left[\frac{K}{\nu_1\nu_2}\left(\frac{1}{\rho_1}+\frac{1}{\rho_2}\right)+\frac{\pi^2\left(n+\frac{1}{2}\right)^2}{h^2}\left(\frac{1}{\nu_1}+\frac{1}{\nu_2}\right)\mp\right.$$

$$\left.\mp\left\{\left[\frac{K}{\nu_1\nu_2}\left(\frac{1}{\rho_1}+\frac{1}{\rho_2}\right)\right]^2+\frac{2K}{\nu_1\nu_2}\left(\frac{1}{\rho_1}-\frac{1}{\rho_2}\right)\left(\frac{1}{\nu_1}-\frac{1}{\nu_2}\right)\frac{\pi^2\left(n+\frac{1}{2}\right)^2}{h^2}+\left[\frac{\pi^2\left(n+\frac{1}{2}\right)}{h^2}\left(\frac{1}{\nu_2}-\frac{1}{\nu_1}\right)\right]^2\right\}^{\frac{1}{2}}\right];$$

and

$$u_2 = -\frac{\frac{p_{1i}}{p_1\mu_1}\frac{p_{2i}}{p_2\mu_2}}{\frac{p_{1i}}{p_1\mu_1}+\frac{p_{2i}}{p_2\mu_2}} N\left\{-\frac{1}{2}(y^2-h^2)-\frac{\frac{1}{\mu_1}-\frac{1}{\mu_2}}{\left(\frac{p_{1i}}{p_1\mu_1}+\frac{p_{2i}}{p_2\mu_2}\right)K}\frac{p_2\mu_2}{p_{2i}}\left[1-\frac{\operatorname{ch}\sqrt{\left(\frac{p_{1i}}{p_1\mu_1}+\frac{p_{2i}}{p_2\mu_2}\right)Ky}}{\operatorname{ch}\sqrt{\left(\frac{p_{1i}}{p_1\mu_1}+\frac{p_{2i}}{p_2\mu_2}\right)Kh}}\right]\right\}+$$

$$+\frac{4}{3}\sum_{n=1}^{\infty}\frac{(-1)^{n+1}}{\left(n+\frac{1}{2}\right)}N\cos\pi\left(n+\frac{1}{2}\right)\frac{y}{h}\left\{\frac{D_5}{D_2}\frac{1}{\gamma_{1n}}e^{-\gamma_{1n}t}-\frac{D_6}{D_4}\frac{1}{\gamma_{2n}}e^{-\gamma_{2n}t}\right\},$$

where

$$D_5 = \frac{\left[-\frac{\rho_{1i}}{\mu_1}\gamma_{1n} + \frac{\rho_{1i}}{\rho_1\mu_1}K + \frac{\pi^2}{h^2}\left(n+\frac{1}{2}\right)^2\right]\frac{1}{\mu_2} + \frac{\rho_{2i}}{\rho_2\mu_2}K\frac{1}{\mu_1}}{-\left(\frac{\rho_{2i}}{\mu_2}+\frac{\rho_{1i}}{\mu_1}\right)\gamma_{1n} + \left(\frac{\rho_{2i}}{\rho_2\mu_2}+\frac{\rho_{1i}}{\rho_1\mu_1}\right)K + 2\frac{\pi^2}{h^2}\left(n+\frac{1}{2}\right)^2} \;;$$

$$D_6 = \frac{\left[-\frac{\rho_{1i}}{\mu_1}\gamma_{2n} + \frac{\rho_{1i}}{\rho_1\mu_1}K + \frac{\pi^2}{h^2}\left(n+\frac{1}{2}\right)^2\right]\frac{1}{\mu_2} + \frac{\rho_{2i}}{\rho_2\mu_2}K\frac{1}{\mu_1}}{-\left(\frac{\rho_{2i}}{\mu_2}+\frac{\rho_{1i}}{\mu_1}\right)\gamma_{2n} + \left(\frac{\rho_{2i}}{\rho_2\mu_2}+\frac{\rho_{1i}}{\rho_1\mu_1}\right)K + 2\frac{\pi^2}{h^2}\left(n+\frac{1}{2}\right)^2} .$$

For $t \to \infty$ we obtain (13.8) and (13.9) for the steady-motion velocities; for $\mu_1 = \mu_2 = \mu$, $\rho_{1i} = \rho_{2i} = \rho$ we have a well-known formula for a viscous incompressible fluid [10]

$$u = -\frac{1}{\mu}\frac{\partial p}{\partial x}\left\{\frac{1}{2}\left(y^2 - h^2\right) + \frac{32h^2}{\nu}\sum_{n=1}^{\infty}\frac{(-1)^n}{\pi^3(2n+1)}e^{-\left(n+\frac{1}{2}\right)^2\frac{\pi^2}{h^2\nu}t}\right\}$$

§ 29. Motion in a Circular Cylindrical Conduit

Let us elucidate the problem of the previous section for a circular cylindrical conduit. Let the 0z axis be along the conduit axis. Let us assume the motion axisymmetric. Then the equations of motion in a cylindrical coordinate system are written as

$$\left.\begin{aligned}\rho_1\frac{\partial u_1}{\partial t} &= f_1\mu_1\left(\frac{\partial^2 u_1}{\partial r^2} + \frac{1}{r}\frac{\partial u_1}{\partial r}\right) + K(u_2 - u_1) - f_1\frac{\partial p}{\partial z} \\ \rho_2\frac{\partial u_2}{\partial t} &= f_2\mu_2\left(\frac{\partial^2 u_2}{\partial r^2} + \frac{1}{r}\frac{\partial u_2}{\partial r}\right) + K(u_1 - u_2) - f_2\frac{\partial p}{\partial z}\end{aligned}\right\} \tag{29.1}$$

The initial and boundary conditions will be

$$\left.\begin{aligned}&\text{for } t = 0 && u_1(r,0) = 0, \quad u_2(r,0) = 0 \\ &\text{for } r = R && (t>0), \quad u_1(r,t) = 0, \quad u_2(r,t) = 0\end{aligned}\right\} \tag{29.2}$$

Let us apply the Laplace–Carson transformation; we then have

$$\left.\begin{aligned}\frac{d^2\bar{u}_1}{dr^2} + \frac{1}{r}\frac{d\bar{u}_1}{dr} - \left(\frac{K}{f_1\mu_1} + \frac{\lambda}{\nu_1}\right)\bar{u}_1 + \frac{K}{f_1\mu_1}\bar{u}_2 &= \frac{N}{\mu_1} \\ \frac{d^2\bar{u}_2}{dr^2} + \frac{1}{r}\frac{d\bar{u}_2}{dr} - \left(\frac{K}{f_2\mu_2} + \frac{\lambda}{\nu_2}\right)\bar{u}_2 + \frac{K}{f_2\mu_2}\bar{u}_1 &= \frac{N}{\mu_2}\end{aligned}\right\} \tag{29.3}$$

where

$$N = \frac{\partial p}{\partial z}.$$

For $r = R$ the conditions (29.2) transform into

$$\left.\begin{aligned}\bar{u}_1 &= 0 \\ \bar{u}_2 &= 0\end{aligned}\right\}. \tag{29.4}$$

Solving the system (29.3) under the boundary conditions (29.4) by the d'Alembert method, and transforming back to the original, we have

$$u_1 = -\frac{\frac{\rho_{1i}}{\rho_1\mu_1}\frac{\rho_{2i}}{\rho_2\mu_2}}{\frac{\rho_{1i}}{\rho_1\mu_1}+\frac{\rho_{2i}}{\rho_2\mu_2}}\,N\left\{-\frac{1}{4}(r^2-R^2)-\frac{\left(\frac{1}{\mu_1}-\frac{1}{\mu_2}\right)\frac{\rho_{2i}}{\rho_2\mu_2}}{\left(\frac{\rho_{1i}}{\rho_1\mu_1}+\frac{\rho_{2i}}{\rho_2\mu_2}\right)K}\left[1-\frac{I_0\left(\sqrt{\left(\frac{\rho_{1i}}{\rho_1\mu_1}+\frac{\rho_{2i}}{\rho_2\mu_2}\right)K}\,r\right)}{I_0\left(\sqrt{\left(\frac{\rho_{1i}}{\rho_1\mu_1}+\frac{\rho_{2i}}{\rho_2\mu_2}\right)K}\,R\right)}\right]\right\}+$$

$$+\sum_{n=1}^{\infty}\frac{4N}{\alpha_n}\frac{J_0\left(\alpha_n\frac{r}{R}\right)}{J_1(\alpha_n)}\left\{\frac{E_1}{E_2}\frac{1}{\gamma_{1n}}e^{-\gamma_{1n}t}-\frac{E_3}{E_4}\frac{1}{\gamma_{2n}}e^{-\gamma_{2n}t}\right\},\tag{29.5}$$

where

$$E_1=\frac{\left[-\frac{\rho_{2i}}{\mu_2}\gamma_{1n}+\frac{\rho_{2i}}{\rho_2\mu_2}K+\frac{\alpha_n^2}{R^2}\right]\frac{1}{\mu_1}+\frac{\rho_{1i}}{\rho_1\mu_1}K\frac{1}{\mu_2}}{-\left(\frac{\rho_{2i}}{\rho_2\mu_2}+\frac{\rho_{1i}}{\rho_1\mu_1}\right)\gamma_{1n}+\left(\frac{\rho_{2i}}{\rho_2\mu_2}+\frac{\rho_{1i}}{\rho_1\mu_1}\right)K+2\frac{\alpha_n^2}{R^2}}\;;$$

$$E_2=-\left(\frac{\rho_{2i}}{\mu_2}+\frac{\rho_{1i}}{\mu_1}\right)+\frac{\left(\frac{\rho_{2i}}{\mu_2}-\frac{\rho_{1i}}{\mu_1}\right)\left[\left(\frac{\rho_{2i}}{\rho_2\mu_2}-\frac{\rho_{1i}}{\rho_1\mu_1}\right)K-\gamma_{1n}\left(\frac{1}{\nu_2}-\frac{1}{\nu_1}\right)\right]}{-\left(\frac{\rho_{2i}}{\mu_2}+\frac{\rho_{1i}}{\mu_1}\right)\gamma_{1n}+\left(\frac{\rho_{2i}}{\rho_2\mu_2}+\frac{\rho_{1i}}{\rho_1\mu_1}\right)K+2\frac{\alpha_n^2}{R^2}}\;;$$

$$E_3=\frac{\left[-\frac{\rho_{2i}}{\mu_2}\gamma_{2n}+\frac{\rho_{2i}}{\rho_2\mu_2}K+\frac{\alpha_n^2}{R^2}\right]\frac{1}{\mu_1}+\frac{\rho_{1i}}{\rho_1\mu_1}K\frac{1}{\mu_2}}{-\left(\frac{\rho_{2i}}{\mu_2}+\frac{\rho_{1i}}{\mu_1}\right)\gamma_{2n}+\left(\frac{\rho_{2i}}{\rho_2\mu_2}+\frac{\rho_{1i}}{\rho_1\mu_1}\right)K+2\frac{\alpha_n^2}{R^2}}\;;$$

$$E_4=-\left(\frac{\rho_{2i}}{\mu_2}+\frac{\rho_{1i}}{\mu_1}\right)+\frac{\left(\frac{\rho_{2i}}{\mu_2}-\frac{\rho_{1i}}{\mu_1}\right)\left[\left(\frac{\rho_{2i}}{\rho_2\mu_2}-\frac{\rho_{1i}}{\rho_1\mu_1}\right)K-\gamma_{2n}\left(\frac{1}{\nu_2}-\frac{1}{\nu_1}\right)\right]}{-\left(\frac{\rho_{2i}}{\mu_2}+\frac{\rho_{1i}}{\mu_1}\right)\gamma_{2n}+\left(\frac{\rho_{2i}}{\rho_2\mu_2}+\frac{\rho_{1i}}{\rho_1\mu_1}\right)K+2\frac{\alpha_n^2}{R^2}}\;;$$

α_n are the roots of the equation $J_0(x)=0$; γ_{1n}, γ_{2n} are determined from the equations

$$I_0\left(\sqrt{c-A_1 b}\,R\right)=0,\qquad I_0\left(\sqrt{c-A_2 b}\,R\right)=0;$$

and A_1, A_2, b, and c are obtained from (27.6) and (27.7).

Analogously we find

$$u_2=\frac{\frac{\rho_{1i}}{\rho_1\mu_1}\frac{\rho_{2i}}{\rho_2\mu_2}}{\frac{\rho_{1i}}{\rho_1\mu_1}+\frac{\rho_{2i}}{\rho_2\mu_2}}\,N\left\{-\frac{1}{4}(r^2-R^2)-\frac{\left(\frac{1}{\mu_2}-\frac{1}{\mu_1}\right)\frac{\rho_{1i}}{\rho_1\mu_1}}{\left(\frac{\rho_{2i}}{\rho_2\mu_2}+\frac{\rho_{1i}}{\rho_1\mu_1}\right)K}\left[1-\frac{I_0\left(\sqrt{\left(\frac{\rho_{1i}}{\rho_1\mu_1}+\frac{\rho_{2i}}{\rho_2\mu_2}\right)K}\,r\right)}{I_0\left(\sqrt{\left(\frac{\rho_{1i}}{\rho_1\mu_1}+\frac{\rho_{2i}}{\rho_2\mu_2}\right)K}\,R\right)}\right]\right\}+$$

$$+\sum_{n=1}^{\infty}\frac{4N}{\alpha_n}\frac{J_0\left(\alpha_n\frac{r}{R}\right)}{J_1(\alpha_n)}\left\{\frac{E_5}{E_2}\frac{1}{\gamma_{1n}}e^{-\gamma_{1n}t}-\frac{E_6}{E_4}\frac{1}{\gamma_{2n}}e^{-\gamma_{2n}t}\right\};\tag{29.6}$$

where

$$E_5 = \frac{\left[-\dfrac{\rho_{1i}}{\mu_1}\gamma_{1n} + \dfrac{\rho_{1i}}{\rho_1\mu_1}K + \dfrac{a_n^2}{R^2} \right]\dfrac{1}{\mu_2} + \dfrac{\rho_{2i}}{\rho_2\mu_2}K\dfrac{1}{\mu_1}}{-\left(\dfrac{\rho_{2i}}{\mu_2} + \dfrac{\rho_{1i}}{\mu_1} \right)\gamma_{1n} + \left(\dfrac{\rho_{2i}}{\rho_2\mu_2} + \dfrac{\rho_{1i}}{\rho_1\mu_1} \right)K + 2\dfrac{a_n^2}{R^2}};$$

$$E_6 = \frac{\left[-\dfrac{\rho_{1i}}{\mu_1}\gamma_{2n} + \dfrac{\rho_{1i}}{\rho_1\mu_1}K + \dfrac{a_n^2}{R^2} \right]\dfrac{1}{\mu_2} + \dfrac{\rho_{2i}}{\rho_2\mu_2}K\dfrac{1}{\mu_1}}{-\left(\dfrac{\rho_{2i}}{\mu_2} + \dfrac{\rho_{1i}}{\mu_1} \right)\gamma_{2n} + \left(\dfrac{\rho_{2i}}{\rho_2\mu_2} + \dfrac{\rho_{1i}}{\rho_1\mu_1} \right)K + 2\dfrac{a_n^2}{R^2}}.$$

In this case, velocity formulas for both the steady and single-phase unsteady motions are obtained under the same assumptions as in the preceding section.

CHAPTER V

STEADY MOTION OF VISCOUS INCOMPRESSIBLE TWO-PHASE MEDIA IN CONDUITS IN THE VARIABLE-POROSITY CASE

It is known that the velocity of a viscous incompressible fluid in Poiseuille flow is distributed identically (parabolically) in any conduit cross section. The velocities of each phase in the motion of viscous incompressible two-phase media are also distributed identically under the assumption of constancy of the reduced densities (see Chapter III), but the shape of the curve is not parabolic. In the general case of steady motion of two-phase media, the reduced densities are variable even when the media are incompressible [3].

A study of the changes in the media velocities in cross sections for parameters given in some other section is of undoubted theoretical and practical interest.

Let us elucidate this problem for the steady motion of a two-phase flow in a plane [30] and circular cylindrical conduit [31].

§ 30. Motion in a Plane Conduit

Let the $0x$ axis be directed along the axis of a plane conduit, and let the $0y$ axis be perpendicular to it. We then have

$$v_1 = 0, \quad v_2 = 0.$$

Let us assume

$$\frac{\partial p}{\partial y} = 0.$$

In this case the system of differential equations of motion for two-phase media (6.2) and (6.3) becomes

$$\rho_1 u_1 \frac{\partial u_1}{\partial x} = -\frac{\rho_1}{\rho_{1i}} \frac{\partial p}{\partial x} + \frac{4}{3} \frac{\partial}{\partial x} \frac{\rho_1}{\rho_{1i}} \mu_1 \frac{\partial u_1}{\partial x} + \frac{\partial}{\partial y} \frac{\rho_1}{\rho_{1i}} \mu_1 \frac{\partial u_1}{\partial y} + K(u_2 - u_1), \tag{30.1}$$

$$\rho_2 u_2 \frac{\partial u_2}{\partial x} = -\frac{\rho_2}{\rho_{2i}} \frac{\partial p}{\partial x} + \frac{4}{3} \frac{\partial}{\partial x} \frac{\rho_2}{\rho_{2i}} \mu_2 \frac{\partial u_2}{\partial x} + \frac{\partial}{\partial y} \frac{\rho_2}{\rho_{2i}} \mu_2 \frac{\partial u_2}{\partial y} + K(u_1 - u_2). \tag{30.2}$$

The continuity equations will be

$$\frac{\partial (\rho_1 u_1)}{\partial x} = 0 \tag{30.3}$$

and

$$\frac{\partial (\rho_2 u_2)}{\partial x} = 0. \tag{30.4}$$

Eliminating $\partial p/\partial x$ from (30.1) and (30.2), we have

$$u_1 \rho_{1i} \frac{\partial u_1}{\partial x} - \rho_{2i} u_2 \frac{\partial u_2}{\partial x} = \frac{4}{3} \frac{\rho_{1i}}{\rho_1} \frac{\partial}{\partial x} \frac{\rho_1}{\rho_{1i}} \mu_1 \frac{\partial u_1}{\partial x} - \frac{4}{3} \frac{\rho_{2i}}{\rho_2} \frac{\partial}{\partial x} \frac{\rho_2}{\rho_{2i}} \mu_2 \frac{\partial u_2}{\partial x} + \frac{\rho_{1i}}{\rho_1} \frac{\partial}{\partial y} \frac{\rho_1}{\rho_{1i}} \mu_1 \frac{\partial u_1}{\partial y} -$$

$$- \frac{\rho_{2i}}{\rho_2} \frac{\partial}{\partial y} \frac{\rho_2}{\rho_{2i}} \mu_2 \frac{\partial u_2}{\partial y} + K \left(\frac{\rho_{1i}}{\rho_1} + \frac{\rho_{2i}}{\rho_2} \right) (u_2 - u_1). \tag{30.5}$$

Let the reduced densities ρ_{10}, ρ_{20} and the media velocities $u_{10}(y)$, $u_{20}(y)$ be known in a given cross section (see § 13). In any other section we have

$$\left. \begin{array}{l} u_1 = u_{10}(y) + u'_1 \\ u_2 = u_{20}(y) + u'_2 \end{array} \right\}, \tag{30.6}$$

$$\rho_1 = \rho_{10} + \rho'_1 , \tag{30.7}$$

$$\rho_2 = \rho_{20} + \rho'_2 , \tag{30.8}$$

where the u'_1, u'_2, ρ'_1, ρ'_2 are very small quantities which vanish at the initial section (i.e., for x = 0) and characterize the changes in the velocities and reduced densities of the media as functions of x and y. Neglecting products of these small quantities, we obtain in place of (30.3), (30.4), (30.5), and (2.1)

$$\rho_{1i} u_{10}(y) \frac{\partial u'_1}{\partial x} - \rho_{2i} u_{20}(y) \frac{\partial u'_2}{\partial x} = \frac{4}{3} \mu_1 \frac{\partial^2 u'_1}{\partial x^2} - \frac{4}{3} \mu_2 \frac{\partial^2 u'_2}{\partial x^2} + \mu_1 \frac{\partial^2 u_{10}(y)}{\partial y^2} - \mu_2 \frac{\partial^2 u_{20}(y)}{\partial y^2} +$$

$$+ \mu_1 \frac{\partial^2 u'_1}{\partial y^2} - \mu_2 \frac{\partial^2 u'_2}{\partial y^2} + \frac{\mu_1}{\rho_{10}} \rho'_1 \frac{\partial^2 u_{10}(y)}{\partial y^2} - \frac{\mu_2}{\rho_{20}} \rho'_2 \frac{\partial^2 u_{20}(y)}{\partial y^2} + \frac{\mu_1}{\rho_{10}} \frac{\partial \overline{\rho}'_1}{\partial y} \frac{\partial u_{10}(y)}{\partial y} -$$

$$- \frac{\mu_2}{\rho_{20}} \frac{\partial \overline{\rho}'_2}{\partial y} \frac{\partial u_{20}(y)}{\partial y} + K \left(\frac{\rho_{1i}}{\rho_{10}} + \frac{\rho_{2i}}{\rho_{20}} \right) [u_{20}(y) - u_{10}(y)] + K \left(\frac{\rho_{1i}}{\rho_{10}} + \frac{\rho_{2i}}{\rho_{20}} \right) (u'_2 - u'_1), \tag{30.9}$$

$$\frac{\partial \rho'_1}{\partial x} u_{10}(y) + \frac{\partial u'_1}{\partial x} \rho_{10} = 0, \tag{30.10}$$

$$\frac{\partial \rho'_2}{\partial x} u_{20}(y) + \frac{\partial u'_2}{\partial x} \rho_{20} = 0, \tag{30.11}$$

$$\frac{\rho'_1}{\rho_{1i}} + \frac{\rho'_2}{\rho_{2i}} = 0. \tag{30.12}$$

Let us add boundary conditions to the equations obtained. For x = 0,

$$u'_1 = u'_2 = 0, \quad \rho'_1 = \rho'_2 = 0, \tag{30.13}$$

and, because of adhesion to the walls, for y = h,

$$u'_1 = u'_2 = 0. \tag{30.14}$$

Our problem now reduces to solving equations (30.9) − (30.12) under the boundary conditions (30.13) and (30.14). To do this, we apply the Laplace–Carson transform. Denoting the transforms of the appropriate originals $\overline{u}_1$, $\overline{u}_2$, $\overline{\rho}_1$, $\overline{\rho}_2$, by u'_1, u'_2, ρ'_1, ρ'_2, we have the following system of ordinary differential equations:

$$\mu_1 \frac{d^2 \overline{u}_1}{dy^2} - \mu_2 \frac{d^2 \overline{u}_2}{dy^2} + \left[\frac{4}{3} \mu_1 \lambda^2 - \rho_{1i} u_{10}(y) \lambda - K \left(\frac{\rho_{1i}}{\rho_{10}} + \frac{\rho_{2i}}{\rho_{20}} \right) \right] \overline{u}_1 - \left[\frac{4}{3} \mu_2 \lambda^2 - \rho_{2i} u_{20}(y) \lambda - K \left(\frac{\rho_{1i}}{\rho_{10}} + \frac{\rho_{2i}}{\rho_{20}} \right) \right] \overline{u}_2 +$$

$$+ \frac{\mu_1}{\rho_{10}} u_{10}(y) \frac{d\bar{\rho_1}}{dy} - \frac{\mu_2}{\rho_{20}} u'_{20}(y) \frac{d\bar{\rho_2}}{dy} - \frac{4}{3} \mu_1 \lambda \left(\frac{\partial u'_1}{\partial x} \right)_{x=0} + \frac{4}{3} \mu_2 \lambda \left(\frac{\partial u'_2}{\partial x} \right)_{x=0} + \mu_1 u'_{10}(y) - \mu_2 u'_{20}(y) +$$

$$+ \frac{\mu_1}{\rho_{10}} u''_{10}(y)\bar{\rho_1} - \frac{\mu_2}{\rho_{20}} u''_{20}(y)\bar{\rho_2} + K \left(\frac{\rho_{1i}}{\rho_{10}} + \frac{\rho_{2i}}{\rho_{20}} \right) [u_{20}(y) - u_{10}(y)] = 0, \tag{30.15}$$

$$\bar{\rho_1} + \frac{\rho_{10}}{u_{10}(y)} \bar{u_1} = 0, \tag{30.16}$$

$$\bar{\rho_2} + \frac{\rho_{20}}{u_{20}(y)} \bar{u_2} = 0, \tag{30.17}$$

$$\bar{\rho_2} + \frac{\rho_{2i}}{\rho_{1i}} \bar{\rho_1} = 0, \tag{30.18}$$

where λ is the transformation parameter, and the boundary conditions (30.14) transform for $y = \pm h$ into

$$\bar{u_1} = \bar{u_2} = 0. \tag{30.19}$$

It is readily noted that

$$\bar{u_1} = - \frac{\rho_{20}}{\rho_{2i}} \frac{\rho_{1i}}{\rho_{10}} \frac{u_{10}(y)}{u_{20}(y)} \bar{u_2} \tag{30.20}$$

from (30.16), (30.17), and (30.18). Taking account of (30.16), (30.17), and (30.20), we rewrite (30.15) as

$$\left[\mu_1 + \mu_2 \frac{\rho_{2i}}{\rho_{20}} \frac{\rho_{10}}{\rho_{1i}} \frac{u_{20}(y)}{u_{10}(y)} \right] \frac{d^2 \bar{u_1}}{dy^2} - \frac{1}{u_{10}(y)} \left[\mu_1 u'_{10}(y) + \mu_2 u'_{20}(y) \frac{\rho_{2i}}{\rho_{20}} \frac{\rho_{10}}{\rho_{1i}} \right] \frac{d\bar{u_1}}{dy} +$$

$$+ \left\{ \frac{4}{3} \left[\mu_1 + \frac{\rho_{2i}}{\rho_{20}} \frac{\rho_{10}}{\rho_{1i}} \frac{u_{20}(y)}{u_{10}(y)} \mu_2 \right] \lambda^2 - \left[\rho_{1i} u_{10}(y) + \frac{\rho_{2i}^2}{\rho_{20}} \frac{\rho_{10}}{\rho_{1i}} \frac{u_{20}^2(y)}{u_{10}(y)} \right] \lambda - \left[K \left(\frac{\rho_{1i}}{\rho_{10}} + \frac{\rho_{2i}}{\rho_{20}} \right) + \right. \right.$$

$$+ K \left(\frac{\rho_{1i}}{\rho_{10}} + \frac{\rho_{2i}}{\rho_{20}} \right) \frac{\rho_{2i}}{\rho_{20}} \frac{\rho_{10}}{\rho_{1i}} \frac{u_{20}(y)}{u_{10}(y)} + \mu_1 \frac{u''_{10}(y)}{u_{10}(y)} + \mu_2 \frac{u''_{20}(y)}{u_{20}(y)} \frac{\rho_{2i}}{\rho_{20}} \frac{\rho_{10}}{\rho_{1i}} - \mu_1 \left(\frac{u'_{10}(y)}{u_{10}(y)} \right)^2 - \mu_2 \frac{u''_{20}(y) u'_{20}(y)}{u_{10}^2(y)} \frac{\rho_{2i}}{\rho_{20}} \frac{\rho_{10}}{\rho_{1i}} \left. \right] \right\} \bar{u_1} =$$

$$= \frac{4}{3} \left[\mu_1 \left(\frac{\partial u'_1}{\partial x} \right)_{x=0} - \mu_2 \left(\frac{\partial u'_2}{\partial x} \right)_{x=0} \right] \lambda + \mu_2 u'_{20}(y) - \mu_1 u'_{10}(y) + K \left(\frac{\rho_{1i}}{\rho_{10}} + \frac{\rho_{2i}}{\rho_{20}} \right) [u_{10}(y) - u_{20}(y)]. \tag{30.21}$$

To simplify the solution of the problem, we replace the initial velocities $u_{10}(y)$ and $u_{20}(y)$ in (30.21) by the mean discharge velocities of the media U_1 and U_2. Then (30.21) becomes

$$\frac{d^2 \bar{u_1}}{dy^2} - A\bar{u_1} = B, \tag{30.22}$$

where

$$A = \frac{-\dfrac{4}{3} \left[\mu_1 + \dfrac{\rho_{2i}}{\rho_{20}} \dfrac{\rho_{10}}{\rho_{1i}} \dfrac{U_2}{U_1} \mu_2 \right] \lambda^2 + \left[\rho_{1i} U_1 + \dfrac{\rho_{2i}^2}{\rho_{20}} \dfrac{\rho_{10}}{\rho_{1i}} \dfrac{U_2^2}{U_1} \right] \lambda + L}{\mu_1 + \mu_2 \dfrac{\rho_{2i}}{\rho_{20}} \dfrac{\rho_{10}}{\rho_{1i}} \dfrac{U_2}{U_1}};$$

$$L = K \left(\frac{\rho_{1i}}{\rho_{10}} + \frac{\rho_{2i}}{\rho_{20}} \right) \left(1 + \frac{\rho_{2i}}{\rho_{20}} \frac{\rho_{10}}{\rho_{1i}} \frac{U_2}{U_1} \right);$$

$$B = \frac{\dfrac{4}{3} \left[\mu_1 \left(\dfrac{\partial u'_1}{\partial x} \right)_{x=0} - \mu_2 \left(\dfrac{\partial u'_2}{\partial x} \right)_{x=0} \right] \lambda + K \left(\dfrac{\rho_{1i}}{\rho_{10}} + \dfrac{\rho_{2i}}{\rho_{20}} \right)(U_1 - U_2)}{\mu_1 + \mu_2 \dfrac{\rho_{2i}}{\rho_{20}} \dfrac{\rho_{10}}{\rho_{1i}} \dfrac{U_2}{U_1}}.$$

The coefficients A and B are constants. Therefore, the general solution of (30.22) under the boundary conditions (30.19) will be

$$\bar{u}_1 = -\frac{B}{A}\left(1 - \frac{\operatorname{ch}\sqrt{A}y}{\operatorname{ch}\sqrt{A}h}\right),$$
(30.23)

where

$$A = -a\lambda^2 + b\lambda + c, \quad \left(a = \frac{4}{3}\right) \left.\begin{array}{c} \\ \\ \end{array}\right\};$$
$$B = s\lambda + r$$
(30.24)

$$b = \frac{\rho_{1i}U_1 + \dfrac{\rho_{10}}{\rho_{1i}}\dfrac{\rho_{2i}^2}{\rho_{20}}\dfrac{U_2}{U_1}}{\mu_1 + \mu_2\dfrac{\rho_{10}}{\rho_{1i}}\dfrac{\rho_{2i}}{\rho_{20}}\dfrac{U_2}{U_1}};$$

$$c = \frac{K\left[\dfrac{\rho_{1i}}{\rho_{10}} + \dfrac{\rho_{2i}}{\rho_{20}}\dfrac{U_2}{U_1} + \left(\dfrac{\rho_{2i}}{\rho_{20}}\right)^2\dfrac{\rho_{10}}{\rho_{1i}}\dfrac{U_2}{U_1} + \dfrac{\rho_{2i}}{\rho_{20}}\right]}{\mu_1 + \mu_2\dfrac{\rho_{10}}{\rho_{1i}}\dfrac{\rho_{2i}}{\rho_{20}}\dfrac{U_2}{U_1}};$$
(30.25)

$$s = \frac{4}{3}\frac{\mu_1\left(\dfrac{\partial u_1'}{\partial x}\right)_{x=0} - \mu_2\left(\dfrac{\partial u_2'}{\partial x}\right)_{x=0}}{\mu_1 + \mu_2\dfrac{\rho_{10}}{\rho_{1i}}\dfrac{\rho_{2i}}{\rho_{20}}\dfrac{U_2}{U_1}};$$

$$r = \frac{K\left(\dfrac{\rho_{1i}}{\rho_{10}} + \dfrac{\rho_{2i}}{\rho_{20}}\right)(U_1 - U_2)}{\mu_1 + \mu_2\dfrac{\rho_{10}}{\rho_{1i}}\dfrac{\rho_{2i}}{\rho_{20}}\dfrac{U_2}{U_1}}.$$

In this notation (30.23) is rewritten as

$$\bar{u}_1 = \frac{s\lambda + r}{a\lambda^2 - b\lambda - c}\left(1 - \frac{\operatorname{ch}\sqrt{c + b\lambda - a\lambda^2}\,y}{\operatorname{ch}\sqrt{c + b\lambda - a\lambda^2}\,h}\right).$$
(30.26)

Let us invert from the transform $\bar{u}_1$ to the original u_1' by means of the formula

$$u' = \frac{1}{2\pi i}\int_{\sigma - i\infty}^{\sigma + i\infty} e^{\lambda x}\frac{s\lambda + r}{a\lambda^2 - b\lambda - c}\left(1 - \frac{\operatorname{ch}\sqrt{c + b\lambda - a\lambda^2}\,y}{\operatorname{ch}\sqrt{c + b\lambda - a\lambda^2}\,h}\right)\frac{d\lambda}{\lambda}.$$
(30.27)

Since the expression

$$\frac{s\lambda + r}{a\lambda^2 - b\lambda - c}\left(1 - \frac{\operatorname{ch}\sqrt{c + b\lambda - a\lambda^2}\,y}{\operatorname{ch}\sqrt{c + b\lambda - a\lambda^2}\,h}\right)\frac{d\lambda}{\lambda}$$

has only simple poles

$$\lambda = \lambda_1 = 0,$$
$$\lambda = \lambda_2 = \alpha > 0,$$
$$\lambda_m = \gamma_m = -\frac{b + \sqrt{b^2 + 4a\left[c + \dfrac{\pi^2}{h^2}\left(\dfrac{2m+1}{2}\right)^2\right]}}{2a},$$

then

$$\frac{s\lambda + r}{a\lambda^2 - b\lambda - c}\left(1 - \frac{\text{ch}\sqrt{c + b\lambda - a\lambda^2}y}{\text{ch}\sqrt{c + b\lambda - a\lambda^2 h}}\right)\frac{1}{\lambda} = \frac{c_0}{\lambda} + \sum_{m=2}^{\infty}\frac{c_{m+2}}{\lambda - \gamma_m}. \qquad (30.28)$$

Let us find the residues

$$c_0 = -\frac{r}{c}\left(1 - \frac{\text{ch}\sqrt{cy}}{\text{ch}\sqrt{ch}}\right),$$

$$c_{m+2} = \frac{2\left(\gamma_m s + r\right)\text{ch}\sqrt{c + b\gamma_m - a\gamma_m^2}\,y}{\gamma_m\sqrt{c + b\gamma_m - a\gamma_m^2 h}\,(b - 2a\gamma_m)\,\text{sh}\sqrt{c + b\gamma_m - a\gamma_m^2 h}}. \qquad (30.29)$$

This latter may be represented as

$$c_{m+2} = (-1)^{m+1}\frac{2\left(\gamma_m s + r\right)\cos\frac{2m+1}{2}\frac{\pi}{h}y}{\frac{2m+1}{2}\pi h\,(b - 2a\gamma_m)\,\gamma_m}. \qquad (30.30)$$

On the basis of (30.28) – (30.30) we have

$$u_1' = -\frac{r}{c}\left(1 - \frac{\text{ch}\sqrt{cy}}{\text{ch}\sqrt{ch}}\right) + \sum_{m=2}^{\infty}(-1)^{m+1}\frac{2\left(\gamma_m s + r\right)\cos\frac{2m+1}{2}\frac{\pi}{h}y}{\frac{2m+1}{2}\pi\,(b - 2a\gamma_m)\,h\gamma_m}\,e^{\gamma_m x}. \qquad (30.31)$$

Completely analogously

$$u_2' = -\frac{\rho_{10}}{\rho_{1i}}\frac{\rho_{2i}}{\rho_{20}}\frac{U_{20}(y)}{U_{10}(y)}\left\{-\frac{r}{c}\left(1 - \frac{\text{ch}\sqrt{cy}}{\text{ch}\sqrt{ch}}\right) + \sum_{m=1}^{\infty}(-1)^{m+1}\frac{2\left(\gamma_m s + r\right)\cos\frac{2m+1}{2}\frac{\pi}{h}y}{\frac{2m+1}{2}\pi\,(b - 2a\gamma_m)\,h\gamma_m}\,e^{\gamma_m x}\right\}. \qquad (30.32)$$

Let us note that the relationship (see § 10)

$$\frac{u_1'}{u_2'} = -\frac{\rho_{1i}}{\rho_{10}}\frac{\rho_{20}}{\rho_{2i}}\frac{U_{10}(y)}{U_{20}(y)}$$

results from (30.31) and (30.32).

Now utilizing (30.6), (30.31), and (30.32), the velocity formulas may be written as

$$u_1 = u_{10}(y) - \frac{r}{c}\left(1 - \frac{\text{ch}\sqrt{cy}}{\text{ch}\sqrt{ch}}\right) + \sum_{m=2}^{\infty}(-1)^{m+1}\frac{2\left(\gamma_m s + r\right)\cos\frac{2m+1}{2}\frac{\pi}{h}y}{\frac{2m+1}{2}\pi\,(b - 2a\gamma_m)\,h\gamma_m}\,e^{\gamma_m x}, \qquad (30.33)$$

$$u_2 = u_{20}(y) - \frac{\rho_{10}}{\rho_{1i}}\frac{\rho_{2i}}{\rho_{20}}\frac{u_{20}(y)}{u_{10}(y)}\left\{-\frac{r}{c}\left(1 - \frac{\text{ch}\sqrt{cy}}{\text{ch}\sqrt{ch}}\right) + \sum_{m=2}^{\infty}(-1)^{m+1}\frac{2\left(\gamma_m s + r\right)\cos\frac{2m+1}{2}\frac{\pi}{h}y}{\frac{2m+1}{2}\pi\,(b - 2a\gamma_m)\,h\gamma_m}\,e^{\gamma_m x}\right\}. \qquad (30.34)$$

For $x \to \infty$ we have for (30.33) and (30.34)

$$u_1 = u_{10}(y) - \frac{r}{c}\left(1 - \frac{\text{ch}\sqrt{cy}}{\text{ch}\sqrt{ch}}\right), \qquad (30.35)$$

$$u_2 = u_{20}(y) + \frac{\rho_{10}}{\rho_{1i}}\frac{\rho_{2i}}{\rho_{20}}\frac{u_{20}(y)}{u_{10}(y)}\frac{r}{c}\left(1 - \frac{\text{ch}\sqrt{c}\,y}{\text{ch}\sqrt{ch}}\right). \qquad (30.36)$$

Therefore, the media velocities at infinity are distributed along curves traced by hyperbolic cosines. The second members in (30.35) and (30.36) are corrections to the velocity distribution formulas in § 13.

Let us note that for identical initial velocities of the media the first members yield coincident parabolas and the second members vanish.

From (30.35) and (30.36) it is readily noted that if $U_1 > U_2$, then u_1 decreases while u_2 increases along the conduit, where both velocities tend to different functions. If $U_1 < U_2$, then u_1 increases while u_2 decreases. In both cases u_1 and u_2 tend to different quantities dependent on y.

§ 31. Motion in a Circular Cylindrical Conduit

Let the 0z axis be directed along the axis of the cylindrical conduit, and let us assume that the steady-state motion of the two-phase media is rectilinear and parallel to this axis. We consider the motion axisymmetric. Then

$$v_{1r} = v_{2r} = 0, \quad v_{1\varphi} = v_{2\varphi} = 0,$$

$$\frac{\partial v_{1r}}{\partial \varphi} = \frac{\partial v_{2r}}{\partial \varphi} = 0.$$

We neglect mass forces. In this case we have instead of the system (8.3) and (3.3)

$$
\left.
\begin{aligned}
&\frac{\rho_1}{\rho_{1i}}\frac{\partial p}{\partial r} + \frac{2}{3}\frac{\mu_1}{\rho_{1i}}\frac{\partial \rho_1}{\partial r}\frac{\partial v_{1z}}{\partial z} - \frac{\mu_1}{\rho_{1i}}\frac{\partial \rho_1}{\partial z}\frac{\partial v_{1z}}{\partial r} - \frac{1}{3}\frac{\mu_1\rho_1}{\rho_{1i}}\frac{\partial^2 v_{1z}}{\partial r \partial z} = 0 \\[6pt]
&\rho_1 v_{1z}\frac{\partial v_{1z}}{\partial z} = -\frac{\rho_1}{\rho_{1i}}\frac{\partial p}{\partial z} + \frac{\mu_1}{\rho_{1i}}\frac{\partial \rho_1}{\partial r}\frac{\partial v_{1z}}{\partial r} + \mu_1\frac{\rho_1}{\rho_{1i}}\frac{\partial^2 v_{1z}}{\partial r^2} + \\[6pt]
&+ \frac{4}{3}\frac{\mu_1}{\rho_{1i}}\frac{\partial \rho_1}{\partial r}\frac{\partial v_{1z}}{\partial z} + \frac{4}{3}\mu_1\frac{\rho_1}{\rho_{1i}}\frac{\partial^2 v_{1z}}{\partial z^2} + \mu_1\frac{\rho_1}{\rho_{1i}}\frac{1}{r}\frac{\partial v_{1z}}{\partial r} + K\left(v_{2z} - v_{1z}\right)
\end{aligned}
\right\}
\tag{31.1}
$$

$$
\left.
\begin{aligned}
&\frac{\rho_2}{\rho_{2i}}\frac{\partial p}{\partial r} + \frac{2}{3}\frac{\mu_2}{\rho_{2i}}\frac{\partial \rho_2}{\partial r}\frac{\partial v_{2z}}{\partial z} - \frac{\mu_2}{\rho_{2i}}\frac{\partial \rho_2}{\partial z}\frac{\partial v_{2z}}{\partial r} - \frac{1}{3}\frac{\mu_2\rho_2}{\rho_{2i}}\frac{\partial^2 v_{2z}}{\partial r \partial z} = 0 \\[6pt]
&\rho_2 v_{2z}\frac{\partial v_{2z}}{\partial z} = -\frac{\rho_2}{\rho_{2i}}\frac{\partial p}{\partial z} + \frac{\mu_2}{\rho_{2i}}\frac{\partial \rho_2}{\partial r}\frac{\partial v_{2z}}{\partial r} + \mu_2\frac{\rho_2}{\rho_{2i}}\frac{\partial^2 v_{2z}}{\partial r^2} + \frac{4}{3}\frac{\mu_2}{\rho_{2i}}\frac{\partial \rho_2}{\partial r}\frac{\partial v_{2z}}{\partial r} + \\[6pt]
&+ \frac{4}{3}\mu_2\frac{\rho_2}{\rho_{2i}}\frac{\partial^2 v_{2z}}{\partial z^2} + \mu_2\frac{\rho_2}{\rho_{2i}}\frac{1}{r}\frac{\partial v_{2z}}{\partial r} + K\left(v_{1z} - v_2\right)
\end{aligned}
\right\}
\tag{31.2}
$$

$$\frac{\partial (\rho_1 v_{1z})}{\partial z} = 0, \tag{31.3}$$

$$\frac{\partial (\rho_2 v_{2z})}{\partial z} = 0. \tag{31.4}$$

Let the values of the motion parameters ρ_1, ρ_2, p be constant in some cross section, for example, let them be ρ_{10}, ρ_{20}, p_0, respectively, at z = 0. Then according to § 14, the media velocities will be functions of the radius r, i.e., $v_{1z0}(r)$, $v_{2z0}(r)$.

Hence, it can be said that in any cross section of the cylindrical conduit

$$
\left.
\begin{aligned}
\rho_1 &= \rho_{10} + \rho_1'(r,\,z) \\
\rho_2 &= \rho_{20} + \rho_2'(r,\,z) \\
v_{1z} &= v_{10z}(r) + v_{1z}'(r,\,z) \\
v_{2z} &= v_{20z}(r) + v_{2z}'(r,\,z) \\
p &= p_0 + p'(r,\,z)
\end{aligned}
\right\},
\tag{31.5}
$$

where

$$
\left.
\begin{array}{ll}
\rho_1'(r,\ z), & \rho_2'(r,\ z) \\[2mm]
v_{1z}'(r,\ z), & v_{2z}'(r,\ z)
\end{array}
\right\}
\tag{*}
$$

are very small and vanish at z = 0.

Substituting (31.5) into (31.1) − (31.4) and neglecting products of very small quantities, we have the following system of linearized equations

$$
\left.
\begin{array}{l}
\dfrac{\rho_{10}}{\rho_{1i}}\dfrac{\partial p'}{\partial r} - \dfrac{\mu_1}{\rho_{1i}}\dfrac{\partial \rho_1'}{\partial z}\dfrac{\partial v_{1z0}}{\partial r} - \dfrac{1}{3}\dfrac{\rho_{10}}{\rho_{1i}}\mu_1\dfrac{\partial^2 v_{1z}'}{\partial r \partial z} = 0 \\[4mm]
\rho_{1i}v_{1z0}\dfrac{\partial v_{1z}'}{\partial z} = -\dfrac{\partial p'}{\partial z} + \dfrac{\mu_1}{\rho_{10}}\dfrac{\partial p'}{\partial r}\dfrac{\partial v_{1z0}}{\partial r} + \mu_1\dfrac{\partial^2 v_{1z0}}{\partial r^2} + \\[4mm]
+\,\mu_1\dfrac{\partial^2 v_{1z}'}{\partial r^2} + \dfrac{4}{3}\mu_1\dfrac{\partial^2 v_{1z}'}{\partial z^2} + \mu_1\dfrac{1}{r}\dfrac{\partial v_{1z0}}{\partial r} + \mu_1\dfrac{1}{r}\dfrac{\partial v_{1z}'}{\partial r} + \dfrac{\rho_{1i}}{\rho_{10}}K\left[(v_{2z0} - v_{1z0}) + (v_{2z}' - v_{1z}')\right]
\end{array}
\right\}
\tag{31.6}
$$

$$
\left.
\begin{array}{l}
\dfrac{\rho_{20}}{\rho_{2i}}\dfrac{\partial p'}{\partial r} - \dfrac{\mu_2}{\rho_{2i}}\dfrac{\partial \rho_2'}{\partial z}\dfrac{\partial v_{2z0}}{\partial r} - \dfrac{1}{3}\dfrac{\rho_{20}}{\rho_{2i}}\mu_2\dfrac{\partial^2 v_{2z}'}{\partial r \partial z} = 0 \\[4mm]
\rho_{2i}v_{2z0}\dfrac{\partial v_{2z}'}{\partial z} = -\dfrac{\partial p'}{\partial z} + \dfrac{\mu_2}{\rho_{20}}\dfrac{\partial \rho_2'}{\partial r} + \mu_2\dfrac{\partial^2 v_{2z0}}{\partial r^2} + \mu_2\dfrac{\partial^2 v_{2z}'}{\partial r^2} + \dfrac{4}{3}\mu_2\dfrac{\partial^2 v_{2z}'}{\partial z^2} + \mu_2\dfrac{1}{r}\dfrac{\partial v_{2z0}}{\partial r} + \\[4mm]
+\,\mu_2\dfrac{1}{r}\dfrac{\partial v_{2z}'}{\partial r} + \dfrac{\rho_{2i}}{\rho_{20}}K\left[(v_{1z0} - v_{2z0}) + (v_{1z}' - v_{2z}')\right]
\end{array}
\right\}
\tag{31.7}
$$

$$
v_{1z0}\dfrac{\partial \rho_1'}{\partial z} + \rho_{10}\dfrac{\partial v_{1z}'}{\partial z} = 0,
\tag{31.8}
$$

$$
v_{2z0}\dfrac{\partial \rho_2'}{\partial z} + \rho_{20}\dfrac{\partial v_{2z}'}{\partial z} = 0.
\tag{31.9}
$$

The expression (2.1) becomes

$$
\dfrac{\rho_1'}{\rho_{1i}} + \dfrac{\rho_2'}{\rho_{2i}} = 0.
\tag{31.10}
$$

Considering the derivatives

$$
\dfrac{\partial^2 v_{1z}'}{\partial z^2},\quad \dfrac{\partial^2 v_{2z}'}{\partial z^2},\quad \dfrac{\partial^2 v_{1z}'}{\partial r \partial z},\quad \dfrac{\partial^2 v_{2z}'}{\partial r \partial z}
$$

in (31.6) and (31.7) to be negligibly small, and then replacing the initial velocities $v_{1z2}(r)$ $v_{2z0}(r)$ by the mean volume velocities of the respective media V_1, V_2, we have the following in the abovementioned section:

$$
\rho_{1i}V_1\dfrac{\partial v_{1z}'}{\partial z} = -\dfrac{\partial p'}{\partial z} + \mu_1\dfrac{\partial^2 v_{1z}'}{\partial r^2} + \dfrac{1}{r}\mu_1\dfrac{\partial v_{1z}'}{\partial r} + \dfrac{\rho_{1i}}{\rho_{10}}K\left[(V_2 - V_1) + (v_{2z}' - v_{1z}')\right],
\tag{31.11}
$$

$$
\rho_{2i}V_2\dfrac{\partial v_{2z}'}{\partial z} = -\dfrac{\partial p'}{\partial z} + \mu_2\dfrac{\partial^2 v_{2z}'}{\partial r^2} + \dfrac{1}{r}\mu_2\dfrac{\partial v_{2z}'}{\partial r} + \dfrac{\rho_{2i}}{\rho_{20}}K\left[(V_1 - V_2) + (v_{1z}' - v_{2z}')\right],
\tag{31.12}
$$

$$
\dfrac{\partial p'}{\partial r} = 0.
\tag{31.13}
$$

Therefore, for the solution of our problem, we have a closed system of linearized equations (31.8) – (31.12) with unknown functions v'_{1z}, v'_{2z}, ρ'_1, ρ'_2, p' and boundary conditions for the velocities obtained on the basis of adhesion of the fluid particles to the wall.

Let us solve this system of equations by the Laplace–Carson functional transform method while taking account of the conditions (*). We then have the following system of ordinary differential equations in the transforms $\overline{v}_{1z}$, $\overline{v}_{2z}$, $\overline{\rho}_1$, $\overline{\rho}_2$, $\overline{p}$ of the corresponding originals v'_{1z}, v'_{2z}, ρ_1^1, ρ'_2, p':

$$\mu_1 \frac{d^2 \overline{v}_{1z}}{dr^2} + \mu_1 \frac{1}{r} \frac{d\overline{v}_{1z}}{dr} + \frac{\rho_{1i}}{\rho_{10}} K(V_1 - V_2 + \overline{v}_{2z} - \overline{v}_{1z}) - \rho_{1i} V_1 \lambda \overline{v}_{1z} - \overline{p}\lambda = 0, \tag{31.14}$$

$$\mu_2 \frac{d^2 \overline{v}_{2z}}{dr^2} + \mu_2 \frac{1}{r} \frac{d\overline{v}_{2z}}{dr} + \frac{\rho_{2i}}{\rho_{20}} (V_2 - V_1 + \overline{v}_{1z} - \overline{v}_{2z}) - \rho_{2i} V_2 \lambda \overline{v}_{2z} - \overline{p}\lambda = 0, \tag{31.15}$$

$$\overline{\rho}_1 + \frac{\rho_{10}}{V_1} \overline{v}_{1z} = 0, \tag{31.16}$$

$$\overline{\rho}_2 + \frac{\rho_{20}}{V_2} \overline{v}_{2z} = 0, \tag{31.17}$$

$$\overline{\rho}_1 + \frac{\rho_{1i}}{\rho_{2i}} \overline{\rho}_2 = 0, \tag{31.18}$$

where λ is the transformation parameter.

The boundary conditions are also transformed; for r = R

$$\overline{v}_{1z} = \overline{v}_{2z} = 0. \tag{31.19}$$

We readily note that (31.16) – (31.18) lead to

$$\overline{v}_{2z} = - \frac{\rho_{10}}{\rho_{1i}} \frac{\rho_{2i}}{\rho_{20}} \frac{V_2}{V_1} \overline{v}_{1z}. \tag{31.20}$$

If $\overline{p}$ is eliminated from (31.14) and (31.15) and (31.20) is taken into account, we have

$$\frac{d^2 \overline{v}_{1z}}{dr^2} + \frac{1}{r} \frac{d\overline{v}_{1z}}{dr} - A^2 \overline{v}_{1z} = B, \tag{31.21}$$

where

$$\left.\begin{array}{c} A^2 = \dfrac{\left(\rho_{1i} V_1 + \dfrac{\rho_{2i}^2}{\rho_{20}} \dfrac{\rho_{10}}{\rho_{1i}} \dfrac{V_2^2}{V_1}\right)\lambda + K \dfrac{\rho_{1i}}{\rho_{10}} \dfrac{\rho_{2i}}{\rho_{20}} \left(1 + \dfrac{\rho_{2i}}{\rho_{20}} \dfrac{\rho_{10}}{\rho_{1i}} \dfrac{V_2}{V_1}\right)}{\mu_1 + \mu_2 \dfrac{\rho_{2i}}{\rho_{20}} \dfrac{\rho_{10}}{\rho_{1i}} \dfrac{V_2}{V_1}} \\[2em] B = \dfrac{K \dfrac{\rho_{1i}}{\rho_{10}} \dfrac{\rho_{2i}}{\rho_{20}} (V_1 - V_2)}{\mu_1 + \mu_2 \dfrac{\rho_{2i}}{\rho_{20}} \dfrac{\rho_{10}}{\rho_{1i}} \dfrac{V_2}{V_1}} \end{array}\right\}. \tag{31.22}$$

For boundary conditions (31.19), equation (31.21) yields the solution

$$\overline{v}_{1z} = - \frac{B}{A^2}\left[1 - \frac{I_0(Ar)}{I_0(AR)}\right]. \tag{31.23}$$

Knowing the transform $\bar{v}_{1z}$, we can turn to determining its original $\bar{v}'_{1z}$ by means of the formula

$$\vec{v}_{1z} = -\frac{B}{2\pi i} \int_{\sigma-i\infty}^{\sigma+i\infty} e^{\lambda z} \frac{1}{a\lambda+b}\left[1 - \frac{I_0(r\sqrt{a\lambda+b})}{I_0(R\sqrt{a\lambda+b})}\right]\frac{d\lambda}{\lambda}, \qquad (31.24)$$

where

$$\left.\begin{aligned}
a &= \frac{p_{1i}V_1 + \dfrac{p_{10}}{p_{1i}}\dfrac{p_{2i}^2}{p_{2i}}\dfrac{V_3}{V_1}}{\mu_1+\mu_2 \dfrac{p_{10}}{p_{1i}}\dfrac{p_{2i}}{p_{20}}\dfrac{V_2}{V_1}} \\[2em]
b &= \frac{K\dfrac{p_{1i}}{p_{10}}\dfrac{p_{2i}}{p_{20}}\left(1-\dfrac{p_{10}}{p_{1i}}\dfrac{p_{2i}}{p_{20}}\dfrac{V_2}{V_1}\right)}{\mu_1+\mu_2\dfrac{p_{10}}{p_{1i}}\dfrac{p_{2i}}{p_{20}}\dfrac{V_2}{V_1}}
\end{aligned}\right\} \qquad (31.25)$$

The integrand may be represented as

$$e^{\lambda z}\frac{1}{a\lambda+b}\left[1 - \frac{I_0(r\sqrt{a\lambda+b})}{I_0(R\sqrt{a\lambda+b})}\right]\frac{1}{\lambda} = \frac{c_0}{\lambda-\lambda_1} + \frac{c_1}{\lambda-\lambda_2} + \sum_{m=1}^{\infty}\frac{c_{m+1}}{\lambda-\lambda_m}; \qquad (31.26)$$

where λ_1, λ_2, λ_m are poles and c_0, c_1, c_{m+1} are residues.

It is easy to compute that

$$\lambda_1 = 0, \quad \lambda_2 = -\frac{b}{a},$$

$$\lambda_m = \gamma_m = -\frac{b+\dfrac{l_m^2}{R^2}}{a},$$

where the l_m are successive roots of

$$J_0(R\sqrt{a\lambda+b}) = 0$$

(the Bessel function). For the residues we also have

$$c_0 = \frac{1}{b}\left[1 - \frac{I_0(\sqrt{b}\,r)}{I_0(\sqrt{b}\,R)}\right],$$

$$c_1 = 0,$$

$$c_{m+1} = \frac{2J_0\left(l_m\dfrac{r}{R}\right)}{al_m\gamma_m J_1(l_m)}.$$

Now the integral (31.24) may easily be evaluated. Hence, we have for v'_{1z}

$$v'_{1z} = -B\left\{\frac{1}{b}\left[1 - \frac{I_0(\sqrt{b}\,r)}{I_0(\sqrt{b}\,R)}\right] + \frac{2}{a}\sum_{m=1}^{\infty}\frac{J_0\left(l_m\dfrac{r}{R}\right)}{l_m\gamma_m J_1(l_m)}e^{\gamma_m z}\right\}. \qquad (31.27)$$

Completely analogously we can obtain for v'_{2z}

$$v'_{2z} = \frac{\rho_{2i}}{\rho_{20}} \frac{\rho_{10}}{\rho_{1i}} \frac{V_2}{V_1} B \left\{ \frac{1}{b} \left[1 - \frac{I_0(\sqrt{b}\,r)}{I_0(\sqrt{b}\,R)} \right] + \frac{2}{a} \sum_{m=1}^{\infty} \frac{J_0\left(l_m \frac{r}{R} \right)}{l_m \gamma_m J_1(l_m)} e^{\gamma_m z} \right\}. \tag{31.28}$$

From (31.16) and (31.17) we have

$$\rho'_1 = \frac{\rho_{10}\,B}{V_1} \left\{ \frac{1}{b} \left[1 - \frac{I_0(\sqrt{b}\,r)}{I_0(\sqrt{b}\,R)} \right] + \frac{2}{a} \sum_{m=1}^{\infty} \frac{J_0\left(l_m \frac{r}{R} \right)}{l_m \gamma_m J_1(l_m)} e^{\gamma_m z} \right\}. \tag{31.29}$$

$$\rho'_2 = - \frac{\rho_{10}}{V_1} \frac{\rho_{2i}}{\rho_{1i}} B \left\{ \frac{1}{b} \left[1 - \frac{I_0(\sqrt{b}\,r)}{I_0(\sqrt{b}\,R)} \right] + \frac{2}{a} \sum_{m=1}^{\infty} \frac{J_0\left(l_m \frac{r}{R} \right)}{l_m \gamma_m J_1(l_m)} e^{\gamma_m z} \right\}. \tag{31.30}$$

We therefore have the following formulas for the velocities and reduced densities

$$v_{1z} = v_{1z0} - B \left\{ \frac{1}{b} \left[1 - \frac{I_0(\sqrt{b}\,r)}{I_0(\sqrt{b}\,R)} \right] + \frac{2}{a} \sum_{m=1}^{\infty} \frac{J_0\left(l_m \frac{r}{R} \right)}{l_m \gamma_m J_1(l_m)} e^{\gamma_m z} \right\}, \tag{31.31}$$

$$v_{2z} = v_{2z0} + \frac{\rho_{2i}}{\rho_{20}} \frac{\rho_{10}}{\rho_{1i}} \frac{V_2}{V_1} B \left\{ \frac{1}{b} \left[1 - \frac{I_0(\sqrt{b}\,r)}{I_0(\sqrt{b}\,R)} \right] + \frac{2}{a} \sum_{m=1}^{\infty} \frac{J_0\left(l_m \frac{r}{R} \right)}{l_m \gamma_m J_1(l_m)} e^{\gamma_m z} \right\}, \tag{31.32}$$

$$\rho_1 = \rho_{10} + \frac{\rho_{10}\,B}{V_1} \left\{ \frac{1}{b} \left[1 - \frac{I_0(\sqrt{b}\,r)}{I_0(\sqrt{b}\,R)} \right] + \frac{2}{a} \sum_{m=1}^{\infty} \frac{J_0\left(l_m \frac{r}{R} \right)}{l_m \gamma_m J_1(l_m)} e^{\gamma_m z} \right\}, \tag{31.33}$$

$$\rho_2 = \rho_{20} - \frac{\rho_{10}}{V_1} \frac{\rho_{2i}}{\rho_{1i}} B \left\{ \frac{1}{b} \left[1 - \frac{I_0(\sqrt{b}\,r)}{I_0(\sqrt{b}\,R)} \right] + \frac{2}{a} \sum_{m=1}^{\infty} \frac{J_0\left(l_m \frac{r}{R} \right)}{l_m \gamma_m J_1(l_m)} e^{\gamma_m z} \right\}. \tag{31.34}$$

For $z \rightarrow \infty$ the formulas presented above evidently become

$$v_{1z} = v_{1z0} - \frac{B}{b} \left[1 - \frac{I_0(\sqrt{b}\,r)}{I_0(\sqrt{b}\,R)} \right], \tag{31.35}$$

$$v_{2z} = v_{2z0} + \frac{\rho_{2i}}{\rho_{20}} \frac{\rho_{10}}{\rho_{1i}} \frac{V_2}{V_1} \frac{B}{b} \left[1 - \frac{I_0(\sqrt{b}\,r)}{I_0(\sqrt{b}\,R)} \right], \tag{31.36}$$

$$\rho_1 = \rho_{10} + \frac{\rho_{10}\,B}{V_1 b} \left[1 - \frac{I_0(\sqrt{b}\,r)}{I_0(\sqrt{b}\,R)} \right], \tag{31.37}$$

$$\rho_2 = \rho_{20} - \frac{\rho_{10}}{V_1} \frac{\rho_{2i}}{\rho_{1i}} \frac{B}{b} \left[1 - \frac{I_0(\sqrt{b}\,r)}{I_0(\sqrt{b}\,R)} \right]. \tag{31.38}$$

Let us note that both the media velocities and the reduced densities are distributed according to curves pictured by Bessel functions over the cross section of a circular cylindrical conduit.

At infinity the velocities of media with identical motion parameters will be equivalent and take a form analogous to the velocity of a single-phase fluid.

Let us determine the pressure drop. To do this, we multiply (31.11) by ρ_{10}/ρ_{1i} and (31.12) by ρ_{20}/ρ_{2i}; the results obtained are then combined. Taking account of (2.1), we obtain

$$\rho_{10}\left(V_1 - \frac{\rho_{2i}}{\rho_{1i}}\frac{V_2^2}{V_1}\right)\frac{\partial v_{1z}'}{\partial z} = -\frac{\partial p'}{\partial z} + \frac{\rho_{10}}{\rho_{1i}}\left(\mu_1 - \mu_2\frac{V_2}{V_1}\right)\frac{\partial^2 v_{1z}'}{\partial r^2} + \frac{\rho_{10}}{\rho_{1i}}\left(\mu_1 - \mu_2\frac{V_2}{V_1}\right)\frac{1}{r}\frac{\partial v_{1z}'}{\partial r}. \tag{31.39}$$

Both sides of (31.39) are then multiplied by $2\pi r\,\frac{\rho_{10}+\rho_1}{\rho_{1i}}\,dr$, and neglecting very small quantities we integrate between 0 and R. Then taking account of (31.13), we have

$$\rho_{10}\left(V_1 - \frac{\rho_{2i}}{\rho_{1i}}\frac{V_2^2}{V_1}\right)\frac{\partial}{\partial z}\int_0^R v_{1z}'\,r\,dr = -\frac{R^2}{2}\frac{\partial p}{\partial z} +$$

$$+ \left(\frac{\rho_{10}}{\rho_{1i}}\right)\left(\mu_1 - \mu_2\frac{V_2}{V_1}\right)\left[R\left(\frac{\partial v_{1z}}{\partial r}\right)_{r=R} + \left(v_{1z}'\right)_{r=0}\right] + \left(\frac{\rho_{10}}{\rho_{1i}}\right)\left(\mu_1 - \mu_2\frac{V_2}{V_1}\right)\left(v_{1z}'\right)_{r=0}.$$

Evaluating the expressions

$$\left.\begin{array}{c}\left(v_{1z}'\right)_{r=0}, \quad \left(\dfrac{\partial v_{1z}'}{\partial r}\right)_{r=R} \\[2em] \displaystyle\int_0^R \dfrac{\partial v_{1z}'}{\partial r}\,r\,dr, \quad \int_0^R v_{1z}'\,r\,dr\end{array}\right\} \tag{31.40}$$

with the aid of recursion formulas for the Bessel functions, and then solving (31.40) for $\partial p/\partial z$, we have for the pressure drop along the conduit

$$\frac{\partial p}{\partial z} = -2B\left\{\frac{1}{R^2}\frac{\rho_{10}}{\rho_{1i}}\left(\mu_1 - \mu_2\frac{V_2}{V_1}\right)\left[\frac{R}{\sqrt{b}}\frac{I_1(\sqrt{b}R)}{I_0(\sqrt{b}R)} + \frac{2}{a}\sum_{m=1}^{\infty}\frac{1}{\gamma_m}e^{-\gamma_m z}\right] - \frac{\rho_{10}}{a}\left(V_1 - \frac{\rho_{2i}}{\rho_{1i}}\frac{V_2^2}{V_1}\right)\sum_{m=1}^{\infty}\frac{1}{l_m}e^{-\gamma_m z}\right\}. \tag{31.41}$$

Let us note that (31.41) for $z \to \infty$ becomes

$$\frac{\partial p}{\partial z} = -\frac{2B}{R\sqrt{b}}\frac{\rho_{10}}{\rho_{1i}}\left(\mu_1 - \mu_2\frac{V_2}{V_1}\right)\frac{I_1(\sqrt{b}R)}{I_0(\sqrt{b}R)}. \tag{31.42}$$

An analysis of the two cases for which $V_1 > V_2$ and $V_1 < V_2$ showed that the media velocities vary along the conduit so that the greater velocity decreases while the lesser increases, the values tending, respectively, to (31.35) and (31.36), which are independent of z.

It is interesting to study the nature of the change in porosities f_1 and f_2. The porosity of a medium having a large initial mean velocity will decrease. Hence, both the reduced densities tend to different values independent of z.

§ 32. Horizontal Motion in a Plane Diffusor

Let a diffusor be formed by two planes making a small angle φ. Let us use cylindrical coordinates and let the r axis be directed along the angle bisector between the planes (Fig. 9). Let us find the laws which govern the motions of the individual phases and the changes in their porosities along the flow by considering the motion to be plane-parallel and steady, and the effect of the mass forces to be negligible [32].

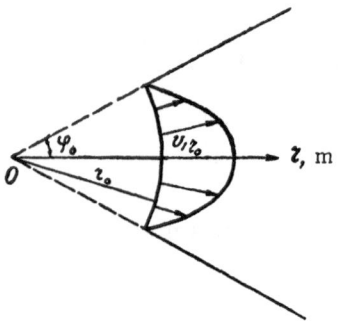

Fig. 9

Let us assume that the motion is strictly radial, i.e, $v_{1\varphi} \approx v_{2\varphi} \approx 0$. By virtue of these assumptions, the equations of motion of a two-phase viscous medium (8.3) are written in simplified form

$$\rho_1 v_{1r} \frac{\partial v_{1r}}{\partial r} = -f_1 \frac{\partial p}{\partial r} + f_1 \mu_1 \left[\frac{4}{3} \frac{\partial^2 v_{1r}}{\partial r^2} + \frac{4}{3} \frac{1}{r} \frac{\partial v_{1r}}{\partial r} \right.$$
$$\left. - \frac{4}{3} \frac{1}{r^2} v_{1r} \right] + \frac{1}{r^2} f_1 \mu_1 \frac{\partial^2 v_{1r}}{\partial \varphi^2} + \left[\frac{4}{3} \frac{\partial v_{1r}}{\partial r} \right.$$
$$\left. - \frac{2}{3} \frac{v_{1r}}{r} \right] \mu_1 \frac{\partial f_1}{\partial r} + \frac{1}{r^2} \mu_1 \frac{\partial v_{1r}}{\partial \varphi} \frac{\partial f_1}{\partial \varphi} + K (v_{2r} - v_{1r})$$

$$\rho_2 v_{2r} \frac{\partial v_{2r}}{\partial r} = -f_2 \frac{\partial p}{\partial r} + f_2 \mu_2 \left[\frac{4}{3} \frac{\partial^2 v_{2r}}{\partial r^2} + \frac{4}{3} \frac{1}{r} \frac{\partial v_{2r}}{\partial r} \right.$$
$$\left. - \frac{4}{3} \frac{1}{r^2} v_{2r} \right] + \frac{1}{r^2} f_2 \mu_2 \frac{\partial^2 v_{2r}}{\partial \varphi^2} + \left[\frac{4}{3} \frac{\partial v_{2r}}{\partial r} - \frac{2}{3} \frac{1}{r} v_{2r} \right] \mu_2 \frac{\partial f_2}{\partial r} + \frac{1}{r^2} \mu_2 \frac{\partial v_{2r}}{\partial \varphi} \frac{\partial f_2}{\partial \varphi} + K (v_{1r} - v_{2r})$$

$$\left. \right\} \quad (32.1)$$

and the continuity equations as

$$v_{1r} \frac{\partial \rho_1}{\partial r} + \rho_1 \frac{\partial v_{1r}}{\partial r} + \frac{1}{r} \rho_1 v_{1r} = 0,$$

$$v_{2r} \frac{\partial \rho_2}{\partial r} + \rho_2 \frac{\partial v_{2r}}{\partial r} + \frac{1}{r} \rho_2 v_{2r} = 0.$$

The boundary conditions will be

for $\varphi = \pm \varphi_0$ (adhesion conditions)
$$v_{1r} = v_{2r} = 0$$
for $r = r_0$
$$v_{1r} = v_{1r0}, \quad v_{2r} = v_{2r0}$$

$$\left. \right\} ; \quad (32.2)$$

where v_{1r0}, v_{2r0} (the values of the velocities at the diffusor entrance) are functions of φ satisfying the adhesion conditions on the boundaries, i.e., for $\varphi = \pm \varphi_0$

$$v_{1r0} = v_{2r0} = 0.$$

Let us eliminate the pressure from the first two equations of the system (32.1) and let us integrate the continuity equations under the boundary conditions (32.2). Introducing the new variables

$$u_1 = v_{1r} \frac{r}{r_0}, \quad (32.3)$$

$$u_2 = v_{2r} \frac{r}{r_0}, \quad (32.4)$$

and then replacing $\ln (r/r_0)$ by ξ, we obtain

$$\left[\rho_{1i} u_1 \left(\frac{\partial u_1}{\partial \xi} - u_1 \right) - \rho_{2i} u_2 \left(\frac{\partial u_1}{\partial \xi} - u_2 \right) \right] r_0 = \frac{4}{3} \mu_1 \left(\frac{\partial^2 u_1}{\partial \xi^2} - 2 \frac{\partial u_1}{\partial \xi} \right) - \frac{4}{3} \mu_2 \left(\frac{\partial^2 u_2}{\partial \xi^2} - 2 \frac{\partial u_2}{\partial \xi} \right) +$$
$$+ \mu_1 \frac{\partial^2 u_1}{\partial \varphi^2} - \mu_2 \frac{\partial^2 u_2}{\partial \varphi^2} + \left[\frac{4}{3} \frac{\partial u_1}{\partial \xi} - 2u_1 \right] \frac{\mu_1}{f_1} \frac{\partial f_1}{\partial \xi} -$$
$$- \left[\frac{4}{3} \frac{\partial u_2}{\partial \xi} - 2u_2 \right] \frac{\mu_2}{f_2} \frac{\partial f_2}{\partial \xi} + \frac{\mu_1}{f_1} \frac{\partial u_1}{\partial \varphi} \frac{\partial f_1}{\partial \varphi} - \frac{\mu_2}{f_2} \frac{\partial u_2}{\partial \varphi} \frac{\partial f_2}{\partial \varphi} + K \left(\frac{1}{f_{11}} + \frac{1}{f_{22}} \right) r_0^2 (u_2 - u_1), \quad (32.5)$$

where

$$\left.\begin{array}{l} f_1\, u_1 = f_{10}\, u_{10} \\ f_2\, u_2 = f_{20}\, u_{20} \\ f_{11} = f_1\, e^{-2\xi} \\ f_{22} = f_2\, e^{-2\xi} \end{array}\right\} \tag{32.6}$$

We apply the small-perturbation method to linearize the system (32.5), (32.6). Let the values of the reduced densities ρ_{10}, ρ_{20}, the porosities f_{10}, f_{20}, and the velocities $u_{10}(\varphi)$, $u_{20}(\varphi)$ be known at a given section. We then have at any other section

$$\left.\begin{array}{l} \rho_1 = \rho_{10} + \rho_1^1 \\ \rho_2 = \rho_{20} + \rho_2^1 \end{array}\right\} \tag{32.7}$$

$$\left.\begin{array}{l} u_1 = u_{10}(\varphi) + u_1' \\ u_2 = u_{20}(\varphi) + u_2' \end{array}\right\} \tag{32.8}$$

$$\left.\begin{array}{ll} f_1 = f_{10} + f_1', & f_{11} = f_{11}^0 + f_{11}' \\ f_2 = f_{20} + f_2', & f_{22} = f_{22}^0 + f_{22}' \end{array}\right\} \tag{32.9}$$

where

$$f_{10} = \frac{\rho_{10}}{\rho_{1i}}; \quad f_{20} = \frac{\rho_{20}}{\rho_{2i}};$$

$$f_1' = \frac{\rho_1'}{\rho_{1i}}; \quad f_2' = \frac{\rho_2'}{\rho_{2i}};$$

and u_1', u_2', ρ_1', ρ_2' are very small quantities which vanish in the initial section and characterize changes in the velocities and reduced densities, and are, in turn, functions of φ, ξ.

Substituting (32.7)−(32.9) into the system (32.5), (32.6), and neglecting products of small quantities, we obtain

$$\left.\begin{array}{l} \mu_1\, \dfrac{\partial^2 u_1'}{\partial \varphi^2} - \mu_2\, \dfrac{\partial^2 u_2'}{\partial \varphi^2} + \mu_1\, u_{10}'\, \dfrac{1}{f_{10}}\, \dfrac{\partial f_1'}{\partial \varphi} - \mu_2\, u_{20}'\, \dfrac{1}{f_{20}}\, \dfrac{\partial f_2'}{\partial \varphi} + \dfrac{4}{3}\left(\mu_1\, \dfrac{\partial^2 u_1'}{\partial \xi^2} - \mu_2\, \dfrac{\partial^2 u_2'}{\partial \xi^2} \right) - \\[2mm] -\left[\left(\dfrac{8}{3}\mu_1 + \rho_{1i}\, u_{10}\, r_0 \right)\dfrac{\partial u_1'}{\partial \xi} - \left(\dfrac{8}{3}\mu_2 + \rho_{2i}\, u_{20}\, r_0 \right)\dfrac{\partial u_2'}{\partial \xi} \right] - \left(2u_{10}\, \dfrac{\mu_1}{f_{10}}\, \dfrac{\partial f_1'}{\partial \xi} - 2u_{20}\, \dfrac{\mu_2}{f_{20}}\, \dfrac{\partial f_2'}{\partial \xi} \right) + \\[2mm] + 2\left(\rho_{1i} u_{10} r_0 u_1' - \rho_{2i} u_{20} r_0 u_2' \right) + K\left(\dfrac{1}{f_{11}^0} + \dfrac{1}{f_{22}^0} \right) r_0^2 \left(u_2' - u_1' \right) + \left(\mu_1\, u_{10}'' - \mu_2\, u_{20}'' \right) + \\[2mm] + \left(\rho_{1i} u_{10}^2 - \rho_{2i} u_{20}^2 \right) r_0 + K\left(\dfrac{1}{f_{11}^0} + \dfrac{1}{f_{22}^0} \right) r_0^2 \left(u_{20} - u_{10} \right) = 0 \\[4mm] \qquad\qquad f_1'\, u_{10} + f_{10}\, u_1' = 0 \\[2mm] \qquad\qquad f_2'\, u_{20} + f_{20}\, u_2' = 0 \\[2mm] \qquad\qquad\qquad f_1' + f_2' = 0 \end{array}\right\} \tag{32.10}$$

where u_{10}', u_{20}' are the first and u_{10}'', u_{20}'' the second derivatives of $u_{10}(\varphi)$ and $u_{20}(\varphi)$ with respect to φ.

For the new variables the boundary conditions will be

for $\xi = 0$
$$u_1' = u_2' = \rho_1' = \rho_2' = f_1' = f_2' = 0, \tag{32.11}$$

for $\varphi = \pm \varphi_0$
$$u_1' = u_2' = 0. \tag{32.12}$$

Our problem has thus been reduced to the solution of the system (32.10) under the boundary conditions (32.11) and (32.12). We apply the Laplace–Carson transform for the solution. Denoting the transforms of the variables u_1', u_2', f_1', f_2' by $\bar{u}_1$, $\bar{u}_2$, $\bar{f}_1$, $\bar{f}_2$, we obtain the following system of equations:

$$\mu_1 \frac{d^2\bar{u}_1}{d\varphi^2} - \mu_2 \frac{d^2\bar{u}_2}{d\varphi^2} + \left\{ \left[\frac{4}{3}\mu_1\lambda^2 - \left(\frac{8}{3}\mu_1 + \rho_{1l}u_{10}r_0 \right)\lambda - K\left(\frac{1}{f_{10}} + \frac{1}{f_{20}} \right)r_0^2 + 2\rho_{1l}u_{10}r_0 \right]\bar{u}_1 - \right.$$

$$- \left[\frac{4}{3}\mu_2\lambda^2 - \left(\frac{8}{3}\mu_2 + \rho_{2l}u_{20}r_0 \right)\lambda - K\left(\frac{1}{f_{10}} + \frac{1}{f_{20}} \right)r_0^2 + 2\rho_{2l}u_{20}r_0 \right]\bar{u}_2 \bigg\} - \left[2u_{10}\mu_1 \frac{1}{f_{10}}\lambda\bar{f}_1 - \right.$$

$$- 2u_{20}\mu_2 \frac{1}{f_{20}}\lambda\bar{f}_2 \bigg] + \left[\mu_1 u_{10}' \frac{1}{f_{10}} \frac{d\bar{f}_1}{d\varphi} - \mu_2 u_{20}' \frac{1}{f_{20}} \frac{d\bar{f}_2}{d\varphi} \right] = \frac{4}{3}\left[\mu_1 \left(\frac{\partial u_1'}{\partial\xi} \right)_{\xi=0} - \right.$$

$$- \mu_2 \left(\frac{\partial u_2'}{\partial\xi} \right)_{\xi=0} \bigg] \lambda + \left(\mu_2 \ddot{u}_{20} - \mu_1 \ddot{u}_{10} \right) + \left(\rho_{2l}u_{20}^2 - \rho_{1l}u_{10}^2 \right)r_0 + K\left(\frac{1}{f_{10}} + \frac{1}{f_{20}} \right)r_0^2\left(u_{20} - u_{10} \right), \quad (32.13)$$

$$\bar{f}_1 + \frac{f_{10}}{u_{10}}\bar{u}_1 = 0, \tag{32.14}$$

$$\bar{f}_2 + \frac{f_{20}}{u_{20}}\bar{u}_2 = 0, \tag{32.15}$$

$$\bar{f}_1 + \bar{f}_2 = 0, \tag{32.16}$$

where λ is the transformation parameter.

For $\varphi = \varphi_0$ the boundary conditions (32.12) go over into

$$\bar{u}_1 = \bar{u}_2 = 0. \tag{32.17}$$

From the system (32.14) − (32.16) we readily find that

$$\bar{u}_2 = -\frac{f_{10}}{f_{20}}\frac{u_{20}}{u_{10}}\bar{u}_1, \tag{32.18}$$

$$\bar{f}_1 = -\frac{f_{10}}{u_{10}}\bar{u}_1, \tag{32.19}$$

$$\bar{f}_2 = \frac{f_{20}}{u_2^0}\bar{u}_2. \tag{32.20}$$

Eliminating the parameters $\bar{u}_2$, $\bar{f}_1$, $\bar{f}_2$ from the system (32.13), (32.18), (32.19), and (32.20), we write a differential equation for $\bar{u}_1$:

$$\left(\mu_1 + \mu_2 \frac{f_{10}u_{20}}{f_{20}u_{10}} \right)\frac{d^2\bar{u}_1}{d\varphi^2} - \frac{1}{u_{10}}\left(\mu_1 u_{10}' + 2\mu_2 \frac{f_{10}u_{20}}{f_{20}u_{10}}u_{10}' - \mu_2 \frac{f_{10}}{f_{20}}u_{20}' \right)\frac{d\bar{u}_1}{d\varphi} +$$

$$+\left\{\frac{4}{3}\left(\mu_1+\mu_2\frac{f_{10}}{f_{20}}\frac{u_{20}}{u_{10}}\right)\lambda^2-\left[\left(\frac{2}{3}\mu_1+\rho_{1i}u_{10}r_0\right)+\left(\frac{2}{3}\mu_2+\rho_{2i}u_{20}r_0\right)\frac{f_{10}}{f_{20}}\frac{u_{20}}{u_{10}}\right]\lambda-\right.$$

$$-K\left(\frac{1}{f_{10}}+\frac{1}{f_{20}}\right)\left(1+\frac{f_{10}}{f_{20}}\frac{u_{20}}{u_{10}}\right)r_0^2-2\left(\rho_{1i}u_{10}+\rho_{2i}u_{20}\frac{f_{10}}{f_{20}}\frac{u_{20}}{u_{10}}\right)r_0+\mu_1\left(\frac{u_{10}'}{u_{10}}\right)^2+\ \cdot$$

$$+\mu_1\frac{f_{10}}{f_{20}}\frac{u_{20}}{u_{10}}\left(\frac{u_{10}'}{u_{10}}\right)^2-\mu_2\frac{f_{10}}{f_{20}}\frac{u_{20}}{u_{10}}\frac{u_{10}'}{u_{10}}-2\mu_2\frac{f_{10}}{f_{20}}\frac{u_{10}'u_{20}'}{u_{10}^2}+\mu_2\frac{f_{10}}{f_{20}}\frac{u_{20}''}{u_{10}}\right\}\bar{u}_1=\frac{4}{3}\left[\mu_1\left(\frac{\partial u_1'}{\partial\xi}\right)_{\xi=0}-\mu_2\left(\frac{\partial u_2'}{\partial\xi}\right)_{\xi=0}\right]\lambda+$$

$$+\left(\mu_2\ u_{20}''-\mu_1\ u_{10}''\right)+\left(\rho_{2i}u_{20}^2-\rho_{1i}u_{10}^2\right)r_0+K\left(\frac{1}{f_{10}}+\frac{1}{f_{20}}\right)r_0^2\left(u_{10}-u_{20}\right).\qquad(32.21)$$

The problem has thus been reduced to the integration of the second-order differential equation (32.21). In order to simplify the solution of this problem, let us replace the initial velocities u_{10} and u_{20} by the mean discharge velocities of the phases U_1 and U_2. Then (32.21) becomes

$$\frac{d^2\bar{u}_1}{d\varphi^2}-A\bar{u}_1=B,\qquad(32.22)$$

where A and B are constants;

$$A=-\left\{\frac{4}{3}\left(\mu_1+\mu_2\frac{f_{10}}{f_{20}}\frac{U_2}{U_1}\right)\lambda^2-\left[\left(\frac{2}{3}\mu_1+\rho_{1i}U_1r_0\right)+\left(\frac{2}{3}\mu_2+\rho_{2i}U_2r_0\right)\frac{f_{10}}{f_{20}}\frac{U_2}{U_1}\right]\lambda-\right.$$

$$\left.-\left[K\left(\frac{1}{f_{10}}+\frac{1}{f_{20}}\right)\left(1+\frac{f_{10}}{f_{20}}\frac{u_2}{u_1}\right)r_0^2-\left(\rho_{1i}U_1+\rho_{2i}U_2\frac{f_{10}}{f_{20}}\frac{U_2}{U_1}\right)r_0\right]\right\}:\left(\mu_1+\mu_2\frac{f_{10}}{f_{20}}\frac{U_2}{U_1}\right);$$

$$B=\left\{\frac{4}{3}\left[\mu_1\left(\frac{\partial u_1'}{\partial\xi}\right)_{\xi=0}-\mu_2\left(\frac{\partial u_2'}{\partial\xi}\right)_{\xi=0}\right]\lambda+\left(\rho_{2i}U_2^2-\rho_{1i}U_1^2\right)r_0+K\left(\frac{1}{f_{10}}+\frac{1}{f_{20}}\right)r_0^2\left(U_1-U_2\right)\right\}:\left(\mu_1+\mu_2\frac{f_{10}}{f_{20}}\frac{U_2}{U_1}\right)$$

Under the boundary conditions (32.17) the general solution of (32.22) will be

$$\bar{u}_1=-\frac{B}{A}\left(1-\frac{\text{ch}\sqrt{A}\,\varphi}{\text{ch}\sqrt{A}\,\varphi_0}\right);\qquad(32.23)$$

where

$$a=\frac{4}{3};$$

$$b=\frac{2}{3r_0}+\frac{\rho_{1i}U_1+\rho_{2i}U_2\frac{f_{10}}{f_{20}}\frac{U_2}{U_1}}{\mu_1+\mu_2\frac{f_{10}}{f_{20}}\frac{U_2}{U_1}};$$

$$c=\frac{K\left(\frac{1}{f_{10}}+\frac{1}{f_{20}}\right)\left(1+\frac{f_{10}}{f_{20}}\frac{U_2}{U_1}\right)-\left(\rho_{1i}U_1+\rho_{2i}U_2\frac{f_{10}}{f_{20}}\frac{U_2}{U_1}\right)\frac{1}{r_0}}{\mu_1+\mu_2\frac{f_{10}}{f_{20}}\frac{U_2}{U_1}};$$

$$m=\frac{4}{3r_0}\frac{\mu_1\left(\frac{\partial u_1'}{\partial\xi}\right)_{\xi=0}-\mu_2\left(\frac{\partial u_2'}{\partial\xi}\right)_{\xi=0}}{\mu_1+\mu_2\frac{f_{10}}{f_{20}}\frac{U_2}{U_1}};$$

$$n = \frac{K\left(\frac{1}{f_{10}} + \frac{1}{f_{20}}\right)(U_1 - U_2) + \left(\rho_{2i}U_2^2 - \rho_{1i}U_1^2\right)\frac{1}{r_0}}{\mu_1 + \mu_2\frac{f_{10}}{f_{20}}\frac{U_2}{U_1}}.$$

Then (32.23) becomes

$$\bar{u}_1 = \frac{\frac{m}{r_0}\lambda + n}{\frac{a}{r_0^2}\lambda^2 - \frac{b}{r_0}\lambda - c}\left[1 - \frac{\mathrm{ch}\sqrt{c + \frac{b\lambda}{r_0} - \frac{a}{r_0^2}\lambda^2}\ \varphi}{\mathrm{ch}\sqrt{c + \frac{b\lambda}{r_0} - \frac{a}{r_0^2}\lambda^2}\ \varphi_0}\right]. \tag{32.24}$$

We go from the transform to the original u_1' by means of the known formula

$$u_1' = \frac{1}{2\pi i}\int_{\sigma - i\infty}^{\sigma + i\infty} e^{\lambda\xi}\frac{\frac{m}{r_0}\lambda + n}{\frac{a}{r_0^2}\lambda^2 - \frac{b}{r_0}\lambda - c}\left[1 - \frac{\mathrm{ch}\sqrt{c + \frac{b\lambda}{r_0} - \frac{a}{r_0^2}\lambda^2}\ \varphi}{\mathrm{ch}\sqrt{c + \frac{b\lambda}{r_0} - \frac{a}{r_0^2}\lambda^2}\ \varphi_0}\right]\frac{d\lambda}{\lambda}. \tag{32.25}$$

The integrand has the simple poles

$$\lambda = \lambda_0 = 0$$

and

$$\lambda_k = \gamma_k r_0 = \frac{b + \sqrt{b^2 + 4a\left[c + \left(\frac{\pi}{r_0\varphi_0}\frac{2k+1}{2}\right)^2\right]}}{2a}r_0;$$

hence, it may be expanded in proper fractions:

$$\frac{\frac{m}{r_0}\lambda + n}{\frac{a}{r_0^2}\lambda^2 - \frac{b}{r_0}\lambda - c}\left[1 - \frac{\mathrm{ch}\sqrt{c + \frac{b}{r_0}\lambda - \frac{a}{r_0^2}\lambda^2}\ \varphi}{\mathrm{ch}\sqrt{c + \frac{b}{r_0}\lambda - \frac{a}{r_0^2}\lambda^2}\ \varphi_0}\right]\frac{1}{\lambda} = \frac{c_0}{\lambda} + \sum_{k=1}^{\infty}\frac{c_k}{\lambda - \gamma_k r_0}. \tag{32.26}$$

Let us find the residues

$$\left.\begin{aligned}c_0 &= -\frac{n}{c}\left[1 - \frac{\mathrm{ch}\sqrt{c}\ r_0\varphi}{\mathrm{ch}\sqrt{c}\ r_0\varphi_0}\right]\\[2em]c_k &= \frac{2(m\gamma_k + n)\mathrm{ch}\sqrt{c + b\gamma_k - a\gamma_k^2}\ r_0\varphi}{r_0\varphi_0\gamma_k\sqrt{c + b\gamma_k - a\gamma_k^2}(b - 2a\gamma_k)\mathrm{sh}\sqrt{c + b\gamma_k - a\gamma_k^2}\ r_0\varphi_0}\end{aligned}\right\} \tag{32.27}$$

The expression obtained may be written as

$$c_k = \frac{(-1)^{k+1}2(m\gamma_k + n)\cos\frac{2k+1}{2}\pi\frac{\varphi}{\varphi_0}}{\gamma_k\frac{2k+1}{2}(b - 2a\gamma_k)\pi r_0\varphi_0}. \tag{32.28}$$

Keeping in mind the known integrals

$$\frac{1}{2\pi i}\int_{\sigma - i\infty}^{\sigma + i\infty} e^{\lambda\xi}\frac{d\lambda}{\lambda} = 1$$

and

$$\frac{1}{2\pi i}\int_{\sigma-i\infty}^{\sigma+i\infty} e^{\lambda\xi}\,\frac{d\lambda}{\lambda-\gamma_k r_0}=e^{\gamma_k r_0\xi},$$

we find from (32.25) – (32.28)

$$u_1'=\frac{n}{c}\left[1-\frac{\operatorname{ch}\sqrt{c}\,r_0\varphi}{\operatorname{ch}\sqrt{c}\,r_0\varphi_0}\right]+\frac{4}{\pi}\sum_{k=1}^{\infty}(-1)^{k+1}\frac{(m\gamma_k+n)\cos\frac{2k+1}{2}\pi\frac{\varphi}{\varphi_0}}{(2k+1)\,\gamma_k\,(b-2a\gamma_k)\,r_0\varphi_0}\,e^{\gamma_k r_0\xi},\qquad(32.29)$$

for the first phase, and

$$u_2'=-\frac{f_{10}}{f_{20}}\frac{u_{20}}{u_{10}}\left[-\frac{n}{c}\left(1-\frac{\operatorname{ch}\sqrt{c}\,r_0\varphi}{\operatorname{ch}\sqrt{c}\,r_0\varphi_0}\right)\right]+\frac{4}{\pi}\sum_{k=1}^{\infty}(-1)^{k+1}\frac{(m\gamma_k+n)\cos\frac{2k+1}{2}\pi\frac{\varphi}{\varphi_0}}{(2k+1)\,\gamma_k\,(b-2a\gamma_k)\,r_0\varphi_0}\,e^{\gamma_k r_0\xi}\quad(32.30)$$

for the second phase. Then, from (32.8) – (32.30);

$$u_1=u_{10}-\frac{n}{c}\left(1-\frac{\operatorname{ch}\sqrt{c}\,r_0\varphi}{\operatorname{ch}\sqrt{c}\,r_0\varphi_0}\right)+\frac{4}{\pi}\sum_{k=1}^{\infty}(-1)^{k+1}\frac{(m\gamma_k+n)\cos\frac{1}{2}(2k+1)\pi\frac{\varphi}{\varphi_0}}{(2k+1)\,\gamma_k\,(b-2a\gamma_k)\,r_0\varphi_0}\,e^{\gamma_k r_0\xi}\qquad(32.31)$$

and

$$u_2=u_{20}-\frac{f_{10}}{f_{20}}\frac{u_{20}}{u_{10}}\left[-\frac{n}{c}\left(1-\frac{\operatorname{ch}\sqrt{c}\,r_0\varphi}{\operatorname{ch}\sqrt{c}\,r_0\varphi_0}\right)+\frac{4}{\pi}\sum_{k=1}^{\infty}(-1)^{k+1}\frac{(m\gamma_k+n)\cos\frac{1}{2}(2k+1)\pi\frac{\varphi}{\varphi_0}}{(2k+1)\,\gamma_{k1}\,(b_1-2a_1\gamma_{k1})\,r_0\varphi_0}\,e^{\gamma_{k1}r_0\xi}\right].\,(32.32)$$

If we turn from a diffusor to the case of plane-parallel motion of two-phase media between two parallel planes (i.e., for $r_0\to\infty$), then by replacing $r_0\varphi$ by y and $r_0\xi$ by x, where y varies between –h and h, and x between 0 and infinity, the formula for the velocity distribution along the flow may be obtained.

Utilizing (32.3), (32.4), (32.31), and (32.32), we finally obtain for the first and second phase velocities

$$v_{1r}=v_{1r_0}-\frac{n}{c}\left(1-\frac{\operatorname{ch}\sqrt{c}\,r_0\varphi}{\operatorname{ch}\sqrt{c}\,r_0\varphi_0}\right)\right]\frac{r_0}{r}+\frac{4}{\pi}\sum_{k=1}^{\infty}(-1)^{k+1}\frac{(m\gamma_k+n)\cos\frac{\pi}{2}(2k+1)\frac{\varphi}{\varphi_0}}{(2k+1)\,\gamma_k\,(b-2a\gamma_k)\,r_0\varphi_0}\left(\frac{r_0}{r}\right)^{1-\gamma_k r_0},\,(32.33)$$

$$v_{2r}=\left[v_{2r_0}+\frac{f_{10}}{f_{20}}\frac{v_{2r_0}}{v_{1r_0}}\frac{n}{c}\left(1-\frac{\operatorname{ch}\sqrt{c}\,r_0\varphi}{\operatorname{ch}\sqrt{c}\,r_0\varphi_0}\right)\right]\frac{r_0}{r}-$$

$$-\frac{f_{10}}{f_{20}}\frac{v_{2r_0}}{v_{1r_0}}\frac{4}{\pi}\sum_{k=1}^{\infty}(-1)^{k+1}\frac{(m\gamma_k+n)\cos\frac{\pi}{2}(2k+1)\frac{\varphi}{\varphi_0}}{(2k+1)\,\gamma_k\,(b-2a\gamma_k)\,r_0\varphi_0}\left(\frac{r_0}{r}\right)^{1-\gamma_k r_0}.\qquad(32.34)$$

Formulas for the porosities of the first and second phases may be found analogously:

$$f_1=f_{10}-\frac{f_{10}}{v_{1r_0}}\left[-\frac{n}{c}\left(1-\frac{\operatorname{ch}\sqrt{c}\,r_0\varphi}{\operatorname{ch}\sqrt{c}\,r_0\varphi_0}\right)\right]+\frac{4}{\pi}\sum_{k=1}^{\infty}(-1)^{k+1}\frac{(m\gamma_k+n)\cos\frac{\pi}{2}(2k+1)\frac{\varphi}{\varphi_0}}{(2k+1)\,\gamma_k\,(b-2a\gamma_k)\,r_0\varphi_0}\left(\frac{r_0}{r}\right)^{-\gamma_k r_0},\,(32.35)$$

$$f_2 = f_{20} + \frac{f_{10}}{v_{1r_0}}\left[-\frac{n}{c}\left(1 - \frac{\mathrm{ch}\,\sqrt{c}\,r_0\varphi}{\mathrm{ch}\,\sqrt{c}\,r_0\varphi_0}\right)\right] + \frac{4}{\pi}\sum_{k=1}^{\infty}(-1)^{k+1}\frac{(m\gamma_k + n)\cos\frac{\pi}{2}(2k+1)\frac{\varphi}{\varphi_0}}{(2k+1)\gamma_k(b - 2a\gamma_k)\,r_0\varphi_0}\left(\frac{r_0}{r}\right)^{-\gamma_k\,r_0}. \quad (32.36)$$

It is seen from (32.33) and (32.34) that the velocities of both media tend to zero, but one more rapidly than the other as a function of the density and initial velocity.

Let us examine the case when both media have the same initial velocity, i.e.,

$$U_1 = U_2 = U$$

and

$$u_{10} = u_{20} = u.$$

As is seen, the rapidity of the velocity drop depends on the second members in the square brackets in (32.33) and (32.34). Since the factor

$$\left(1 - \frac{\mathrm{ch}\,\sqrt{c}\,r_0\varphi}{\mathrm{ch}\,\sqrt{c}\,r_0\varphi_0}\right)$$

is always positive, the sign of these members depends on the signs of n and c. When r_0 is sufficiently large, c has a positive value, and hence the tendency of the velocity to zero in this case depends on the sign of n; for $\rho_{1i} > \rho_{2i}$ it is negative, and for $\rho_{1i} < \rho_{2i}$ positive. Therefore, for

$$U_1 = U_2 = U$$

the velocity of that medium for which the density is greater will tend to zero more slowly. It follows from (32.35) and (32.36) that the porosities of the first and second media tend to different constants.

It should be noted that for sufficiently large r/r_0 (i.e., sufficiently large distance from the initial section), it is possible to limit oneself in (32.33) and (32.34) to the quantities in square brackets. The mass content of the mixture can be determined by means of the formula

$$\gamma = \gamma_0 + \frac{\rho_{1i}}{\rho_{10}^2}\left(\rho_{20} + \rho_{10}\frac{\rho_{2i}}{\rho_{1i}}\right)(f_1 - f_{10}).$$

Substituting the value of f_1 from (32.35) into the resulting formula, we have

$$\gamma = \gamma_0 + \frac{\rho_{1i}}{\rho_{10}^2}\left(\rho_{20} + \rho_{10}\frac{\rho_{2i}}{\rho_{1i}}\right)\frac{f_{10}}{v_{1r_0}}\left[\frac{n}{c}\left(1 - \frac{\mathrm{ch}\,\sqrt{c}\,r_0\varphi}{\mathrm{ch}\,\sqrt{c}\,r_0\varphi_0}\right) + \frac{4}{\pi}\sum_{k=1}^{\infty}(-1)^{k+1} \times\right.$$

$$\left.\times\frac{(m\gamma_k + n)\cos\frac{\pi}{2}(2k+1)\frac{\varphi}{\varphi_0}}{(2k+1)\gamma_k(b - 2a\gamma_k)r_0\varphi_0}\left(\frac{r_0}{r}\right)^{-\gamma_k\,r_0}\right]. \quad (32.37)$$

Hence, for $U_1 = U_2$ the mass content of the heavy phase diminishes as the diffusor cross section increases, while that of the light phase increases.

CHAPTER VI

THE DRAG COEFFICIENT OF A CONDUIT FOR TWO-PHASE MEDIA IN MUTUALLY PENETRATING MOTION

§ 33. Derivation of Conduit Drag Coefficient Formulas for the Motion of Two-Phase Media

The determination of the drag coefficient of conduits of different cross section for the mutually penetrating motion of viscous two-phase media is of great practical interest.

The drag in conduits for two-phase mixtures in turbulent motion has been examined in many works, for example, in [33, 34]. Drag coefficient formulas are usually developed with the aid of experimental results, without which it would be impossible to obtain theoretical deductions.

We show the possibility of a theoretical investigation of the drag coefficient of a conduit for the case of steady laminar motion of mutually penetrating two-phase media.

The drag coefficient of a conduit for the motion of two-phase media is determined from the formula

$$\lambda_{mi} = \frac{|\ f_1\ \tau_{1max} + f_2\ \tau_{2max}\ |}{f_1\ \dfrac{\rho_{1i}\ u_{1mi}^2}{2} + f_2\ \dfrac{\rho_{2i}\ u_{2mi}^2}{2}}\ , \tag{33.1}$$

where u_{1m}, u_{2m} are the mean velocities and

$$\tau_{1max} = \mu_1 \left(\frac{\partial u_1}{\partial r}\right)_{r=R}, \quad \tau_{2max} = \mu_2 \left(\frac{\partial u_2}{\partial r}\right)_{r=R}$$

are the maximum values of the viscous stresses of the media.

As is seen from (33.1), under the conditions $f_1 = 1$, $f_2 = 0$ or $f_1 = 0$, $f_2 = 1$ the formula for λ_{mi} takes the form of the drag coefficient of a conduit with single-phase medium motion.

§ 34. Motion in a Circular Conduit

Let us derive a formula for the drag coefficient of a circular conduit. To do this, let us use the velocity formulas (14.20) and (14.21). Computations have shown that

$$\left| f_1\ \tau_{1max} + f_2\ \tau_{2max} \right| = \left| \frac{\partial p}{\partial x} \right| \frac{R}{2}\ .$$

133

Then after some manipulation (33.1) becomes

$$\lambda_{mi} = \frac{\left|\frac{\partial p}{\partial x}\right| R}{u_{1m}^2 \, \rho_{1i} \left[f_1 + f_2 \, \frac{\rho_{2i}}{\rho_{1i}} \left(\frac{u_{2m}}{u_{1m}}\right)^2 \right]} \ . \tag{34.1}$$

Let G_1 be the mass flow rate per second of the first medium, i.e.,

$$G_1 = g \rho_{1i} \, f_1 \, \pi R^2 u_{1m} \, ,$$

from which

$$\rho_{1i} \, u_{1m} = \frac{G_1}{g f_1 \pi R^2} \ .$$

Substituting the equality obtained into (34.1), we have

$$\lambda_{mi} = \frac{\left|\frac{\partial p}{\partial x}\right| g f_1 \pi R^3}{u_{1m} \, G_1 \left[f_1 + f_2 \, \frac{\rho_{2i}}{\rho_{1i}} \left(\frac{u_{2m}}{u_{1m}}\right)^2 \right]} \ . \tag{34.2}$$

Let us express the mean velocities in this formula in terms of the media discharges as

$$\left. \begin{array}{l} Q_1 = f_1 \, F u_{1m} \\ Q_2 = f_2 \, F u_{2m} \end{array} \right\} , \tag{34.3}$$

where $F = \pi R^2$ is the area of the conduit section.

Taking account of the formulas for Q_1 and Q_2 obtained in § 14, we have from (34.3)

$$u_{1m} = - \frac{2 \frac{\partial p}{\partial x}}{KR \left[\frac{1}{f_1 \mu_1} + \frac{1}{f_2 \mu_2} \right]} \chi_1' \, , \tag{34.4}$$

$$u_{2m} = - \frac{2 \frac{\partial p}{\partial x}}{KR \left[\frac{1}{f_1 \mu_1} + \frac{1}{f_2 \mu_2} \right]} \chi_2' \, ; \tag{34.5}$$

where

$$\chi_1' = \left[\frac{1}{\frac{1}{f_1 \mu_1} + \frac{1}{f_2 \mu_2}} - \frac{1}{\mu_1} \right] \frac{I_1\left(\sqrt{K \left(\frac{1}{f_1 \mu_1} + \frac{1}{f_2 \mu_2} \right)} \, R \right)}{\sqrt{K \left[\frac{1}{f_1 \mu_1} + \frac{1}{f_2 \mu_2} \right]} I_0\left(\sqrt{K \left[\frac{1}{f_1 \mu_1} + \frac{1}{f_2 \mu_2} \right]} \, R \right)} +$$

$$+ \left[\frac{KR^2}{8 f_1 \mu_1 \, f_2 \mu_2} + \frac{1}{\mu_1} - \frac{1}{\frac{1}{f_1 \mu_1} + \frac{1}{f_2 \mu_2}} \right] \frac{R}{2} \ ;$$

$$\chi_2' = \left[\frac{1}{f_1\mu_1 + f_2\mu_2} - \frac{1}{\mu_2} \right] \frac{I_1\left(\sqrt{K\left[\frac{1}{f_1\mu_1} + \frac{1}{f_2\mu_2}\right]R} \right)}{\sqrt{K\left[\frac{1}{f_1\mu_1} + \frac{1}{f_2\mu_2}\right]} I_0\left(\sqrt{K\left[\frac{1}{f_1\mu_1} + \frac{1}{f_2\mu_2}\right]R} \right)} +$$

$$+ \left[\frac{KR^2}{8\,f_1\mu_1 f_2\mu_2} + \frac{1}{\mu_2} - \frac{1}{f_1\mu_1 + f_2\mu_2} \right] \frac{R}{2}.$$

Dividing (34.4) by (34.5), we obtain

$$\frac{u_{2m}}{u_{1m}} = \frac{\chi_2^1}{\chi_1'}. \tag{34.6}$$

Substituting (34.4) and (34.6) into (34.2), we have

$$\lambda_{mi} = \frac{g f_1 K \pi R^4 \left[\frac{1}{f_1\mu_1} + \frac{1}{f_2\mu_2} \right]}{2\chi_1' G_1 \left[f_1 + f_2 \frac{\rho_{2i}}{\rho_{1i}} \left(\frac{\chi_2'}{\chi_1'} \right)^2 \right]}. \tag{34.7}$$

Introducing the notation $\chi_1 = \chi_1'/R$ and $\chi_2 = \chi_2'/R$, let us rewrite this last equation as

$$\lambda_{mi} = \frac{KR^2 g\pi R}{2G_1}\psi; \tag{34.8}$$

where

$$\psi = \frac{1 + \frac{f_2}{f_1}\frac{\mu_2}{\mu_1}}{f_2\chi_1\mu_2 \left[1 + \frac{f_2}{f_1}\frac{\rho_{2i}}{\rho_{1i}}\left(\frac{\chi_2}{\chi_1} \right)^2 \right]}.$$

Let us note that under the conditions $\rho_{1i} = \rho_{2i}$, $\mu_1 = \mu_2$, $f_1 = 1$, $f_2 = 0$, formula (34.8) becomes

$$\lambda = \frac{8g\pi\mu_1 R}{G}, \tag{34.9}$$

where G is the mass flow rate per second of a single-phase medium.

Dividing (34.8) by (34.9) and considering $G_1 = G$, we obtain

$$\frac{\lambda_{mi}}{\lambda} = \frac{1}{16}\frac{KR^2}{\mu_1}\psi, \tag{34.10}$$

whence, by introducing the notation $H = KR^2/\mu_1$, $\mu = \mu_2/\mu_1$, $\rho_1 = \rho_{2i}/\rho_{1i}$, we have

$$\frac{\lambda_{mi}}{\lambda} = \frac{1}{16}H\psi, \tag{34.11}$$

in which

$$\psi = \frac{1 + \frac{f_2}{f_1}\mu}{T_1 \cdot f_2 \left[1 + \frac{f_2}{f_1}\rho_i\left(\frac{T_2}{T_1} \right)^2 \right]};$$

$$T_1 = \left[\frac{\mu}{f_1 + f_2\mu} - \mu \right] \frac{I_1\left(\sqrt{\frac{f_1 + f_2\mu}{f_1 f_2\mu}} H \right)}{\sqrt{\frac{f_1 + f_2\mu}{f_1 f_2\mu}} H \, I_0\left(\sqrt{\frac{f_1 + f_2\mu}{f_1 f_2\mu}} H \right)} + \frac{1}{2}\left[\frac{H}{8 f_1 f_2} + \mu - \frac{\mu}{f_1 + f_2\mu} \right];$$

$$T_2 = \left[\frac{\mu}{f_1 + f_2\mu} - 1 \right] \frac{I_1\left(\sqrt{\frac{f_1 + f_2\mu}{f_1 f_2\mu}} H \right)}{\sqrt{\frac{f_1 + f_2\mu}{f_1 f_2\mu}} H \, I_0\left(\sqrt{\frac{f_1 + f_2\mu}{f_1 f_2\mu}} H \right)} + \frac{1}{2}\left[\frac{H}{8 f_1 f_2} + 1 - \frac{\mu}{f_1 + f_2\mu} \right].$$

It follows from (34.11) that the ratio of the drag coefficient of a two-phase medium to the drag coefficient of a single-phase medium is a function of the dimensionless quantities H, μ, ρ_i, f_1, f_2.

In order to study the change in the drag coefficient of a conduit with two-phase mixture motion as a function of the volume content of one medium if the mass flow rate per second of the second medium remains constant, λ_{mi}/λ was evaluated on an electronic computer for different values of f_2. Results of calculations for large volume contents of the gas phase in a gas–liquid mixture are presented as a graph in Fig. 10. As is seen from the graph, the mixture drag coefficient diminishes as the gas content by volume increases. This phenomenon is explained physically by the fact that the coefficient of viscosity of the mixture, favoring diminution in the conduit drag coefficient, apparently decreases as the gas content by volume in the fluid increases. This agrees with the experimental results of Pavlova, who considered plug-like motion in [34].

§ 35. Motion between Conduits

A formula for the drag coefficient can be derived by the method described in § 34 for the motion of a two-phase medium in the space between conduits (see § 18):

$$\lambda_{mi} = \frac{K(R^2 - \sigma^2)^2 g \pi \left[2\Phi + \frac{R^2 - \sigma^2}{\sigma} \right]}{2 G_1} \xi, \quad \xi = \frac{1 + \frac{f_2}{f_1} \frac{\mu_2}{\mu_1}}{\theta_1 f_2 \mu_2 \left[1 + \frac{f_2}{f_1} \frac{\rho_{2i}}{\rho_{1i}} \left(\frac{\theta_2}{\theta_1} \right)^2 \right]},$$

$$\theta_1 = \left[\frac{1}{\mu_1} - \frac{1}{f_1\mu_1 + f_2\mu_2} \right] \Phi\sigma - \frac{K}{16(f_1\mu_1 + f_2\mu_2)}(R^2 - \sigma^2) + \left(\frac{1}{\mu_1} - \frac{1}{f_1\mu_1 + f_2\mu_2} \right) \frac{R^2 - \sigma^2}{2} -$$
$$- \frac{KR^2}{4(f_1\mu_1 + f_2\mu_2)} \left[R^2 \ln \frac{\sigma}{R} + \frac{R^2 - \sigma^2}{2} \right],$$

$$\theta_2 = \left[\frac{1}{\mu_2} - \frac{1}{f_1\mu_1 + f_2\mu_2} \right] \Phi\sigma - \frac{K}{16(f_1\mu_1 + f_2\mu_2)}(R^2 - \sigma^2) +$$
$$+ \left(\frac{1}{\mu_2} - \frac{1}{f_1\mu_1 + f_2\mu_2} \right) \frac{R^2 - \sigma^2}{2} - \frac{KR^2}{4(f_1\mu_1 + f_2\mu_2)} \times$$
$$\times \left[R^2 \ln \frac{\sigma}{R} + \frac{R^2 - \sigma^2}{2} \right],$$

$$\Phi = \frac{I_1(m\sigma) K_1(mR) - K_1(m\sigma) I_1(mR)}{m \left[I_0(m\sigma) \cdot K_1(mR) + K_0(m\sigma) I_1(mR) \right]}.$$

On the basis of the formulas obtained we investigate the change in drag coefficient (Fig. 11) for the motion of a gas–liquid mixture in the space between conduits of different radii for a constant mass flow of the liquid phase values for

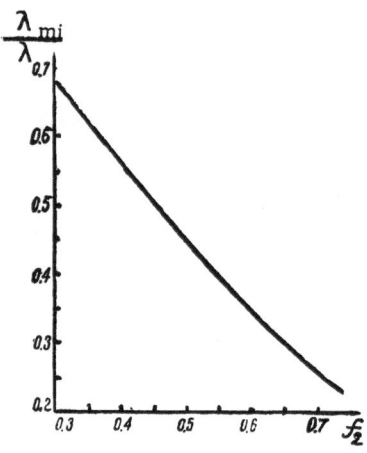

Fig. 10

different σ and the parameter values

$$f_1 = 0.7, \quad f_2 = 0.3, \quad \mu_1 = 0.51 \cdot 10^{-3} \text{ kg} \cdot \text{sec}/\text{m}^2,$$

$$\mu_2 = 1.5310^{-6} \text{ kg} \cdot \text{sec}/\text{m}^2, \quad \rho_{1i} = 80 \text{ kg} \cdot \text{sec}^2/\text{m}^4,$$

$$\rho_{2i} = 0.125 \text{ kg} \cdot \text{sec}^2/\text{m}^4 \; R = 0.07 \text{ m}, \quad K = 0.1 \text{ kg} \cdot \text{sec}/\text{m}^4.$$

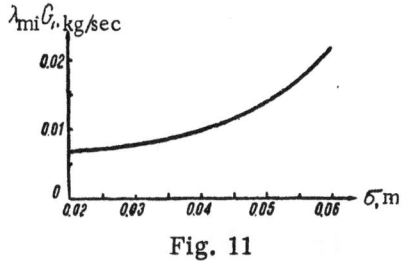

Fig. 11

It follows from Fig. 11 that the drag coefficient λ_{mi} will increase for $G_1 = \text{const}$ as the conduit radius increases.

CHAPTER VII

SOME THEORETICAL AND EXPERIMENTAL INVESTIGATIONS TO DETERMINE THE VALUES OF PHYSICOMECHANICAL CONSTANTS

As has been remarked in Chapter I, each phase of a mixture is considered to be continuous and to move in a porous medium; the pressure is transmitted directly from particle to particle of the phase under consideration. However, this is valid only when the concentration of the suspension is great. At low concentrations it is necessary to consider the fluid as the fundamental medium ("carrier"), and the particles of the suspension as interconnected by means of this medium [35]. Then the terms containing the pressure drop out of the equations of motion.

If the "carrying" fluid is a compressible gas, then the equation of state for the gas "carrier" must be added to the equations of motion [36]. In this case, other physical relationships (between the pressures) must be taken into account in place of the equation of state in order to close the system when the fundamental medium, the "carrier," is a viscous incompressible fluid.

Up to now the investigations have been carried out only under the assumption of high concentrations. Below we determine the interaction coefficients and the provisional viscosity coefficient for such a concentration.

§ 36. Methods of Determining the Interaction Coefficient K

As is seen from the previous chapters, all the formulas for the two-phase flow parameters contain an unknown interaction coefficient K (if the phases are ideal). In the case of viscous gases, there also are viscosity coefficients μ_1 and μ_2 in addition to the interaction coefficient. If the second phase consists of solid particles, the value of the provisional viscosity coefficient μ_2 turns out to be unknown. The coefficients depend on the kind of media in the mixture motion. It is evidently necessary to know the values of these coefficients in order to analyze the motion. Hence, working out the simplest methods of determining them is of practical interest.

Several methods of determining these coefficients exist, utilizing both theoretical formulas and experiments. Shvidler [37] considers the motion of a two-phase flow in a porous medium by using filtration formulas, and gives a method to obtain these coefficients. However, no quantitative value of K is presented in his paper.

The method described in [38] is based on the velocity formulas (13.8) and (13.9). On the basis of these formulas, we obtain after some manipulation

$$\frac{u_{1max} - u_{2max}}{u_{1max}} = \frac{\mu - 1}{\mu - \dfrac{\mu}{f_1 + f_2\mu} + \dfrac{H}{2f_2\mu} \dfrac{\text{ch}\sqrt{\dfrac{\frac{f_1}{\mu} + f_2}{f_1 f_2}H}}{\text{ch}\sqrt{\dfrac{\frac{f_1}{\mu} + f_2}{f_1 f_2}H} - 1}}, \tag{36.1}$$

138

where u_{1max}, u_{2max} are the axial phase velocities; and $\mu = \mu_2/\mu_1$, $H = Kh^2/\mu_1$ are dimensionless quantities.

Therefore, the ratio of the relative velocity to the velocity of the first medium is a function of the dimensionless quantities μ, H, f_1, and f_2.

Let us consider the percentage change in the ratio of the relative velocity to the velocity of the first medium for different values of H and f_1 as a function of μ:

$$\frac{u_{1max} - u_{2max}}{u_{max}} 100\%.$$

The coefficient of viscosity of water is taken as μ_1:

$$\mu f_1 = 1.16 \cdot 10^{-4} \text{ kg} \cdot \text{sec/m}^2.$$

For the case $f_2 = 1/2$ we present a graph (Fig. 12), from which it is seen that the relative velocity as a percentage of the velocity of the first medium is an increasing function of μ. Graphs for other values of f_1, μ, and H may be composed completely analogously.

By knowing the nature of the change $u_{1max} - u_{2max}/u_{max}$ as a function of the dimensionless parameters μ, f_1, and H we may give a method for determining the interaction coefficient K.

The graphs in Fig. 12 have been composed for several values of the parameter H. We determine μ after knowing the viscosity coefficients of the moving media in the mixture. If experimental values of the axial velocities of the media u_{1max} and u_{2max} are known, then by knowing the content of the media by volume we can determine the appropriate value of H on the basis of these graphs. It is then always possible to determine the interaction coefficient K for given μ_1 and H from the equality $H = Kh^2/\mu_1$.

The method of determining K in [38] based on experimental measurements of the axial phase velocities is still more difficult to realize because of the lack of experimental methods of studying these velocities. However, it is easy to reduce the problem of determining the value of the coefficient K to the measurement of the mean flow rate of the phases.

§ 37. Determination of K and μ_2 by Experimental Methods

Experimental investigations to determine the coefficients K and μ_2 were conducted in the Hydrodynamics Laboratory of the Mechanics Institute and Computation Center of the Academy of Sciences of the UzbekSSR, managed by the author, by two methods: by measuring the friction on a plate moving over the surface of a two-phase medium [6], and by measuring the volume discharges of each phase in the flow of two-phase media in a horizontal cylindrical conduit [39].*

The first method reduces to the following: If a motion is produced for which the friction on the conduit wall may be determined by both the theoretical formula and experimentally, then by equating these values for given times t_0 and t_1 we will obtain two equations with two unknowns K and μ_2. By solving these equations, we determine the values of the unknown coefficients.

This method is applied in [6] to the motion considered in § 27. The experimental investigation was conducted in a special hydrodynamic

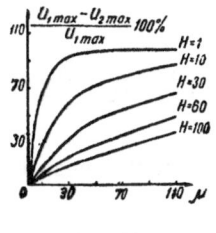

Fig. 12

* Prior to beginning the tests, the experimental apparatus was specially calibrated for pure water flow modes.

closed trough filled with a two-phase fluid (water + rosin) on whose surface a smooth plate was set in motion by a string loaded by a weight. The velocity of the plate's motion was recorded by means of an electronic circuit with a photoresistor and a magneto-electric oscillograph. The friction force was determined by means of the formula

$$M \frac{\partial u}{\partial t} = P - F_f \ ,$$

where M is the mass of the mobile system, P the weight of the load, and F_f the frictional force.

The experimental formula for the frictional force is obtained as

$$F_f = P - M\left(0.272\, e^{-0.15t} - 0.0998\, e^{-1.32t}\right), \tag{37.1}$$

where the mixture temperature corresponded to 21°C and

$$f_1 = 0.75, \quad \mu_1 = 1.024 \cdot 10^{-4} \ \text{kg} \cdot \text{sec/m}^2, \quad \rho_{1i} = 101.7 \ \text{kg} \cdot \text{sec}^2/\text{m}^4,$$

$$\rho_{2i} = 109.1 \ \text{kg} \cdot \text{sec}^2/\text{m}^4.$$

The following quantitative values

$$K = 446 \ \text{kg} \cdot \text{sec/m}^4, \quad \mu_2 = 23 \cdot 10^{-4} \ \text{kg} \cdot \text{sec/m}^2$$

are obtained as a result of comparing the theoretical (31.6) and experimental (37.1) formulas for the frictional force by means of analysis on an electronic computer.

The experimental determination of the coefficients K and μ_2 by measurement of the volume discharges of the phases (the second method) was conducted on a special apparatus consisting of a circular glass conduit 10 mm in diameter and 1.3 m long. The nature of the two-phase flow was photographed, and besides, the mixture flow mode was observed visually. The experiments were conducted with a two-phase mixture (water + rosin) with $f_1 = 0.7$. The following flow parameters were measured: media discharges, pressure drops in the working section of the conduit, and mixture temperature.

Processing of the data showed that the dependence of the media discharges on the pressure drop (Fig. 13) is a straight line in this case [40], which confirms the visually observed laminar flow mode.

We apply (14.22), (14.23) to determine the coefficient μ_2:

$$\mu_2 = \mu_1 \frac{Q_1}{Q_2} + \left(-\frac{\partial p}{\partial x}\right) 0.39 \frac{R^4}{Q_2} \ . \tag{37.2}$$

On the basis of the experimental results, we obtain by using (37.2)

$$\mu_2 = 12.83 \cdot 10^{-4} \ \text{kg} \cdot \text{sec/m}^2.$$

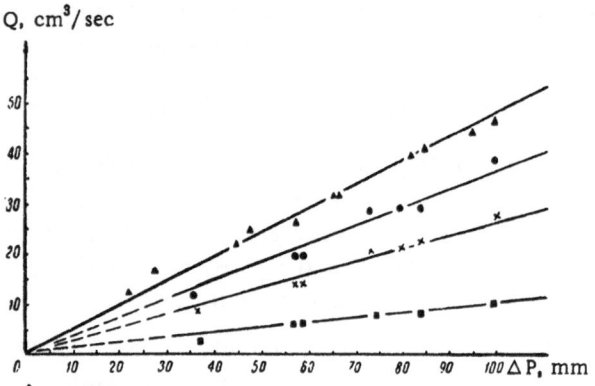

Q, cm³/sec

▲ pure water
● mixture
✕ liquid phase in the mixture
■ solid phase in the mixture

Fig. 13

Furthermore, we used the formula

$$\frac{u_{1m} - u_{2m}}{u_{1m}} = \frac{\mu - 1}{\mu\left(1 - \frac{1}{f_1 + f_2 H}\right) \div \frac{H}{8 f_1 f_2} \dfrac{I_0\sqrt{\dfrac{f_1 + f_2\mu}{f_1 f_2\mu}\,H}}{I_1\sqrt{\dfrac{f_1 + f_2\mu}{f_1 f_2\mu}\,H}}},$$

to determine $H = Kh^2/\mu_1$ and the coefficient K easily from experimental data for u_{1m}, u_{2m}, μ_1, μ_2, and f_1. It has been established on the basis of the tests that

$$K = 483 \ \text{kg} \cdot \text{sec}^2/\text{m}^4.$$

Values of K obtained by these methods differ by 8%, which indicates the correctness of the methods used for the experimental investigation and the accuracy of the tests.

The coefficient of mixture viscosity can be computed from these values of K and μ_2 obtained. This is done as follows. Let a mixture consisting of two phases be subject to the Newton hypothesis (the subscript "mi" refers to the mixture):

$$\tau_{mi} = \mu_{mi}\frac{\partial u_{mi}}{\partial n}.$$

It is known that

$$\tau_{mi} = f_1\mu_1\frac{\partial u_1}{\partial n} + f_2\mu_2\frac{\partial u_2}{\partial n};$$

then

$$\mu_{mi}\frac{\partial u_{mi}}{\partial n} = f_1\mu_1\frac{\partial u_1}{\partial n} + f_2\mu_2\frac{\partial u_2}{\partial n}.$$

Since we have

$$u_{mi} = f_1 u_1 + f_2 u_2,$$

the previous equality is rewritten as

$$\mu_{mi}\left(f_1\frac{\partial u_1}{\partial n} + f_2\frac{\partial u_2}{\partial n}\right) = f_1\mu_1\frac{\partial u_1}{\partial n} + f_2\mu_2\frac{\partial u_2}{\partial n},$$

from which

$$\mu_{mi} = \frac{f_1\mu_1 + f_2\mu_2 M}{f_1 + f_2\mu}, \tag{37.3}$$

where

$$M = \frac{\dfrac{\partial u_2}{\partial n}}{\dfrac{\partial u_1}{\partial n}}.$$

If motion in a circular cylindrical conduit is examined, then by using the formulas in § 14, we can average M over the conduit cross section by using the theorem of the mean. The integral obtained is then evaluated by the Gauss formula.

By using the experimental results for K and μ_2 (for the motion of water + rosin case) and computing the value of μ_{mi} on an electronic computer with the aid of (37.3) for the concentration, we found that our results for $f_2 = 0.3$ agree with the experimental results presented in [41]. This indicates that the theory expounded herein is valid for the flow of mixtures with a considerable concentration, and it is hence impossible to compare (37.3) with the Einstein formula obtained for a mixture with low concentration:

$$\mu_{mi} = \mu_1 (1 + cf),$$

where μ_1 is the fluid viscosity coefficient, f the concentration, and c a coefficient depending on the kind of material.

The experimental results agree with some of the theory elucidated herein which is based on the model of motion of multiphase media given by Rakhmatulin [1].

Popov [42], [43] formulated tests to verify the theory given in [1] in application to problems of a multijet ejector, and he verified the correctness of the multiphase fluid model underlying this theory, in which flow interaction is taken into account.

In subsequent investigations the nature of K and μ_2, the relative phase velocity as a function of various factors, and the heat and mass exchange in multiphase flows taking account of chemical processes should all be studied.

LITERATURE CITED

1. Rakhmatulin, Kh. A., "Principles of the gas dynamics of mutually penetrating media," Prikl. Matem. i Mekh., Vol. 20, No. 2 (1956).
2. Faizullaev, D. F., "On steady motions of viscous incompressible two-phase media," Abstract of Candidate's Dissertation, Tashkent, 1959.
3. Kleiman, Ya. Z., "On the motion of multicomponent media," Abstract of Candidate's Dissertation, Moscow, 1958.
4. Kleiman, Ya. Z., "On steady mixture motion in conduits," Nauchn. Dokl. Vyssh. Shkoly, Fiz.-mat. Nauk, Moscow, No. 4 (1958).
5. Bakhriev, G. B., "On the motion of multicomponent media in conduits," Izv. An TadzhikSSR, Otd. Geol-khim. i tekhn. Nauk, No. 3 (1961).
6. Latipov, K. Sh., "Steady and unsteady motion of two-component media in conduits of variable and constant cross section," Abstract of Candidate's Dissertation, Tashkent, 1964.
7. Stepanov, V. V., Differential Equations, Moscow, Gostekhizdat, 1953.
8. Lavrent'ev, M. A., and Shabat, B. V., Methods of the Theory of Functions of Complex Variables, Moscow, Gostekhizdat, 1951.
9. Faizullaev, D. F., "Poiseuille problem for mutually penetrating motion of two-phase media," Izv. AN UzbekSSR, Ser. tekhn. nauk, No. 3 (1958).
10. Slezkin, N. A., Dynamics of a Viscous Incompressible Fluid, Moscow, Gostekhizdat, 1955.
11. Watson, G. N., Theory of Bessel Functions, Cambridge University Press, 1944.
12. Kotov, Ya. P., Umarov, G. Ya., and Faizullaev, D. F., "On the stationary flow of a conducting medium in the presence of a magnetic field," Izv. AN UzbekSSR, Ser. Fiz.-mat. Nauk, No. 3 (1962).
13. Cowling, T., Magnetohydrodynamics, New York, Interscience, 1957.
14. Latipov, K. Sh., "Motion of a two-phase fluid in an elliptical conduit," Izv. AN UzbekSSR, Ser. tekhn. nauk, No. 4 (1961).
15. Maclachlan, N. W., Theory and Application of Mathieu Functions, New York, Dover, 1947.
16. Leibenzon, L. S., Handbook on Petroleum Mechanics, Pt. I: Hydraulics, Moscow, Gostoptekhizdat, 1931.
17. Kuprin, A. I., and Chen Da-jun, "Some results of investigating the shape of the pulpline cross section of hydraulic transport," Izv. Vuzov, Gornyi Zhur. No. 11 (1960).
18. Faizullaev, D. F., "Generalization of the Poiseuille problem for two-phase media in a circular annular conduit," Izv. AN UzbekSSR, Ser. tekhn. nauk, No. 5 (1958).
19. Faizullaev, D. F., and Narkabulov, O., "On the theory of two-phase media motion in the space between heaters (conduits)," Izv. AN UzbekSSR, Ser tekhn. nauk, No. 2 (1963).
20. Faizullaev, D. F., "Some questions of the steady circular motion of viscous incompressible two-phase media," Izv. AN UzbekSSR, Ser. tekh. nauk, No. 3 (1959).
21. Leonov, A. I., "On the steady flow of a viscous fluid in finite plane and circular conduits," Izv. AN SSSR, OTN, Mekh. i Mashinostr., No. 6 (1962).
22. Nazarii, M. P., "On the steady motion of viscous two-phase media in a circular cylindrical conduit," Izv. AN UzbekSSR, Ser. tekhn. nauk, No. 4 (1965).

144   LITERATURE CITED

23. Targ, S. M., Fundamental Problems of Laminar Flow Theory, Moscow, Gostekhizdat, 1954.

24. Umarov, A. I., "Motion of mutually penetrating viscous incompressible three-phase media between two walls," Voprosy Mekhaniki, No. 2 (1965).

25. Spivakovskii, A. O., Muchnik, V. S., Yufin, A. P., Smoldyrev, A. E., Ofengenden, N. E., Borisenko, L. D., and Trainis, V. V., Hydraulic and Pneumatic Transport in Mines, Moscow, Gosgortekhizdat, 1962.

26. Faizullaev, D. F., "On the question of hydraulic transport of fine-grained material of diverse coarseness," Izv. AN UzbekSSR, Ser. tekhn. nauk, No. 2 (1965).

27. Faizullaev, D. F., and Bagranov, E. A., "Velocity distribution over the cross section in motion of fine-grained materials of diverse coarsness between parallel walls, Doklady AN UzbekSSR, No. 6 (1965).

28. Latipov, K. Sh., "Some problems of the unsteady flow of two-component viscous media," Izv. AN UzbekSSR, Ser. tekhn. nauk, No. 6 (1962).

29. Latipov, K. Sh., "On some problems of unsteady flow of two-component viscous media," Izv. AN UzbekSSR, Ser. tekhn. nauk, No. 4 (1963).

30. Faizullaev, D. F., "On the steady motion of incompressible two-phase media between parallel walls," Izv. AN UzbekSSR, Ser. tekhn. nauk, No. 6 (1961).

31. Umarov, A. I., and Faizullaev, D. F., "On mutually penetrating motion of viscous incompressible two-phase media in a circular cylindrical conduit," Izv. AN UzbekSSR, Ser. tekhn. nauk, No. 4 (1962).

32. Latipov, K. Sh., "Motion of a two-component medium in a plane diffusor," Izv. AN UzbekSSR, Ser. tekhn. nauk, No. 4 (1964).

33. Teletov, S. G., "Hydrodynamics of two-phase fluids," Doctoral Dissertation, Energetics Institute of the Academy of Sciences of the USSR, Moscow, 1948.

34. Pavlova, N. N., "On the question of the drag coefficient in air–water mixture motion in horizontal conduits," Leningrad Institute of Railroad Transport Engineers, Publication No. 185, Leningrad, 1962.

35. Bednarczyk, H., Über die Strömung von Zweiphasengemischen, in Proc. 8th Internat. Astronaut. Congr., Barcelona, Vienna, Springer Verlag, 1958, pp. 47-52.

36. Shul'gin, D. F., "On the theory of the motion of a mixture of a gas and a mechanical suspension," Trudy V. I. Lenin Tashkent State Univ., No. 189, Tashkent, 1961.

37. Shvidler, M. I., "Mutually penetrating motion of immiscible fluids in a porous medium," Izv. AN SSSR, OTN, Mekh. i Mashinostr., No. 1 (1961).

38. Faizullaev, D. F., "On a method of determining the interaction coefficient in the mutually penetrating motion of two-phase media," Doklady AN UzbekSSR, No. 9 (1961).

39. Umarov, A. I., "Theoretical and experimental investigation of the laminar steady motion of multiphase media," Abstract of Candidate's Dissertation, Tashkent, 1965.

40. Umarov, A. I., "Dependence of media discharges on pressure drop in one case of two-phase media motion in a circular conduit," Doklady AN UzbekSSR, No. 12 (1964).

41. Kurgaev, E. F., "On the viscosity of suspensions," Doklady AN SSSR, Vol. 132, No. 2 (1960).

42. Popov, N. N., "New method of analyzing the mixing stage of a multistage ejector," Vestnik Mosk. Univ., No. 3 (1959).

43. Popov, N. N., "On the question of mixing gas flows," Izv. Vuzov, Aviats. Tekh., No. 3 (1960).